Computing for Engineers

Computing for Engineers

ROGER T. FENNER

Ph.D., B.Sc. (Eng.), D.I.C., A.C.G.I.

Lecturer in Mechanical Engineering,
Imperial College of Science and Technology, London

First edition 1974
Reprinted 1978

Published by
THE MACMILLAN PRESS LTD
*London and Basingstoke
Associated companies in Delhi Dublin
Hong Kong Johannesburg Lagos Melbourne
New York Singapore and Tokyo*

ISBN 0 333 15189 5

Printed in Hong Kong

Contents

Contents

7 Structural Analysis and Finite Element Methods 144

8 Further Applications, and Classification of Problems 162

x   Contents

Preface

It is generally acknowledged that digital computation is becoming increasingly important to engineers, and most university courses now include some training in the use of computers at both undergraduate and postgraduate levels. The emphasis is usually on teaching numerical-analysis techniques and a particular computer programming language, with relatively little attention being given to engineering applications. The main purpose of this book is to examine such applications. A wide range of practical problems, taken from several branches of engineering, are examined with the aid of particular case studies. These examples would provide suitable computing projects for teaching purposes, and typical programs and results are presented in each case. It is shown that most engineering problems involve computational techniques that fall into one or more of three main categories. These are the solution of algebraic equations, the solution of differential equations and the general processing of numerical data using, say, statistical methods.

A prior knowledge of the FORTRAN programming language is assumed, although the material is arranged so that the first chapter of applications involves only elementary FORTRAN and avoids the use of subprograms. The importance of thorough program testing, and the advantages of using subprograms are emphasised. The level of numerical analysis and other mathematics used is that normally taught in undergraduate engineering courses. This book is therefore suitable for undergraduate students, and others at an equivalent level. Postgraduates and practising engineers may also find it useful if they are comparatively new to computer methods.

The author wishes to thank his brother, Dr D. N. Fenner, for reading the manuscript and making many helpful suggestions. The assistance of Dr R. A. Knight on the railway-train problem presented in Chapter 4 is gratefully acknowledged, as are the typing services of Miss E. A. Quin.

Imperial College of Science and Technology,   R. T. Fenner
London

1 Introduction

In recent years high-speed electronic digital computers have become increasingly important tools available to the engineer. There are innumerable applications for computers in all branches of engineering and in all types of engineering activity, including research, development, production and management. While it is not necessary for all engineers to become expert computer programmers, it is desirable that they have at least an appreciation of what programming involves and, in particular, how the results obtained are influenced by the fallibility of the programmer.

The main advantages of a digital computer are that it can perform simple arithmetic many millions of times faster than the human brain, and that the reliability of the answer is extremely high. Apart from reducing the cost of calculations, the use of computers has encouraged the application of numerical analysis techniques to problems which are otherwise intractable. A computer cannot solve a problem in the sense of applying intelligent reasoning but it can assist in the exploration of the range of possible solutions. In the design of an engineering component, for example, it cannot define the design criteria, but it can help to predict the consequences of selecting particular component geometry.

The term computer *programming* is not uniquely defined, but for the purposes of this book is taken to mean the complete process of translating a problem into a set of computer instructions which causes the required solution to be produced by the machine. The more restricted activity of writing down the instructions in an appropriate language is termed *coding*, although the final set of instructions is called a program.

The main purpose of this book is to explain and demonstrate the use of the FORTRAN language and various numerical analysis techniques in the application of digital computing to the solution of practical engineering problems. It is intended neither as a FORTRAN manual nor a numerical analysis text, but concentrates on typical applications in the field of engineering science. In this chapter, digital computer hardware and software are briefly reviewed,

1

with particular emphasis on features which affect the user of a high-level language such as FORTRAN.

1.1 Computer Hardware

While it is often convenient to use the term *computer* to describe the equipment used to process a program, in most modern installations it is not a single machine, but a series of machines, each with its own function. Computers vary in size from desk-top units to large installations handling many programs simultaneously and communicating with a large number of remote terminals. The user of a typical general purpose computer normally supplies his program in the form of either a deck of punched cards or punched paper tape. The information is transferred onto magnetic tape or disc by means of an appropriate reading unit, and its further processing is discussed in Section 1.3.

At the heart of the computing hardware is the arithmetic unit which performs the numerical manipulations specified by the program. This unit is continually in communication with the various storage devices. The most important of these is the fast *core store* which is composed of a large number of *words* or storage registers, each of which may contain, for example, a number being used in the calculation. The amount of fast-store hardware readily available to the user is limited, typically to between about ten thousand and a hundred thousand words, depending on the size of the computer. Further storage capacity is available in the form of *backing stores*, usually magnetic tapes or discs, but these are relatively wasteful of computing time.

The results of the calculation are written onto magnetic tape and in due course recorded on paper by a line printer. Despite the fact that these machines print complete lines of characters simultaneously, they can often cause bottle-necks in the whole process if excessive amounts of output are produced by the program.

The stored-program method of operation, in which the program of instructions is contained in the fast store, is common to virtually all modern electronic digital computers. Only by this means can the potential speed of the electronic circuitry be realised, by allowing long sequences of arithmetic operations to be performed without manual intervention. Many large computers are designed for multi-programming, whereby more than one program is stored and executed simultaneously. While only one instruction is executed at a time, the various programs are suitably intermixed. The object is to maximise the use of the input and output devices, fast store and arithmetic unit. Only part of the fast store is available for each program, so that one that requires a large amount of storage will generally be given low priority. Similarly, a program which produces a large amount of output may be delayed at the printing stage.

Many computers are also of the multi-access type, where the user can work at a terminal remote from the machine but connected by a telephone line. There may be many of these terminals in use at any one time, in addition to the normal central submission of cards and tapes. The user communicates with the computer via the terminal teleprinter and keyboard. In this way, a more conversational mode of operation is obtainable. It is also possible to feed digital information from electronic measuring devices directly into a computer, for the purposes of data analysis and control of the process being monitored.

The usual media for input and output communications have already been mentioned. Often a pictorial representation of results is desirable, and most computers have plotting facilities for preparing graphs and diagrams. If greater interaction is required between computer and user, then oscilloscope displays are available and are widely used in, for example, computer-aided design. The user outlines a component or structure on a screen with the aid of, say, a 'light pen'. The computer is programmed to store this geometric information and to project any view of the three-dimensional body. Using appropriate programs for stress analysis, the designer can make alterations to the geometry until the design criteria are met, whereupon the computer can prepare the necessary engineering drawings.

1.2   Representation of Numbers and Characters in a Computer

One aspect of computer hardware deserves rather more detailed examination, as it affects the writing of programs and the interpretation of the results. This is the method of representing and storing numbers and characters in a computer. The majority of modern computers work in binary arithmetic, where all numbers have a radix of two, instead of ten as in the decimal system. Thus, for example, 45_{10} and $0 \cdot 8125_{10}$ are identical to 101101_2 and $0 \cdot 1101_2$ respectively, where the suffixes specify the radixes. It is important to note, however, that decimal numbers do not necessarily have exact binary equivalents. For example, $0 \cdot 1_{10}$ is $0 \cdot 0001100110011_2$ with infinite repetition of the four-digit group 0011. The main advantage of the binary system is that only two digits are required, and these may be represented by the 'on' and 'off' conditions of electrical signals.

1.2.1 Storage of numbers   As already indicated, the fast store of a computer is made up of many words in which numbers may be stored. Each word consists of between about 24 and 60 binary digits or 'bits'. The number of bits is usually fixed for a particular computer, and the current trend is towards more bits per word.

It is important to distinguish between the two main types of numbers used in computations, namely INTEGER and REAL in FORTRAN terminology. Integer numbers can have magnitudes from zero in steps of one up to some maximum value imposed by the finite word length. Their main use is as counters in repetitive loops, and as subscripts. Suppose that a particular computer has a word length of 36 bits. Allowing one bit to represent the sign of an INTEGER number leaves 35 bits to contain its magnitude. Thus the largest storable integer is $2^{35} - 1$. While it is possible to perform arithmetic operations using INTEGER numbers, it should be realised that the result is truncated to the nearest lower integer. For example, the result of dividing 5 by 6 is zero.

REAL numbers, on the other hand, are used in most calculations and are usually stored in floating-point form, as a binary mantissa (fractional part) and exponent. Again for a 36 bit word, the mantissa might occupy 28 bits and the exponent 8 bits. Therefore, allowing one sign bit for each, the mantissa is limited to 27 bits, so that the precision of a stored REAL number is limited to about eight decimal digits. Similarly, the maximum size of exponent is $2^7 - 1$ or 127. Since this is a binary exponent, the largest permissible decimal exponent is about 38. Therefore, any number calculated which has a magnitude greater than about 10^{38} cannot be stored, a condition known as *overflow*. Similarly, a number with a magnitude less than about 10^{-38}, with the exception of zero itself, causes *underflow*.

1.2.2 Roundoff errors In later chapters considerable attention is given to numerical errors in computed results. The most serious of these are usually associated with the particular methods of analysis used. Errors may also arise, however, from the arithmetic operations in the computer, and are known as roundoff errors. These are due to the fact that the result of an arithmetic operation using REAL numbers cannot necessarily be represented exactly by a finite number of digits. As an illustration in decimal arithmetic, consider the operation $1 \cdot 0 \div 3 \cdot 0$, which gives $0 \cdot 33333333$ to eight significant figures. If this result is now multiplied by $3 \cdot 0$, then $0 \cdot 99999999$ is obtained rather than one. In some types of computations such errors may be cumulative, so that roundoff can significantly affect results. Additional roundoff errors are introduced by the conversions between decimal and binary, and the lack of exact equivalence.

Although the word length in most modern computers is sufficient to avoid significant roundoff errors in the majority of computations, it is usually possible to increase precision by linking pairs of words together to form single storage registers. In FORTRAN, this facility is known as DOUBLE PRECISION. The disadvantages are that the storage space is effectively halved and the computing time greatly increased.

The lack of precision in REAL arithmetic has some important influences on the writing of programs. For example, it is generally inadvisable to rely on two numbers being exactly equal, when usually what is required is that they should differ by no more than some suitably small prescribed tolerance.

1.2.3 Storage of characters In addition to numbers, it is also possible to store characters, including letters, decimal digits and any other coding symbols. Each character has its own binary equivalent in a particular computer, involving either six or eight bits. The main application for this facility is in the storage and manipulation of alphanumeric titles and captions.

1.3   Computer Software

The physical components of a computer have already been discussed under the heading of hardware. Another very important part of a modern computing system is the software, which controls the processing of the users' programs. A detailed discussion of software would be inappropriate here, if only because one of its main functions is to ensure that the normal user need know very little about the particular computing system. Programs are normally written in a high-level language, such as FORTRAN, which is designed for use in particular applications, and which is, as far as possible, independent of the computer used.

A program written in a language such as FORTRAN is known as a *source* program. Before it can be used to control the operation of the computer, it must be translated into a far more detailed low-level form known as the *object* program. This specifies the numerous elementary machine instructions involved in even the simplest FORTRAN statements. The part of the software that translates the source program into the object program is known as the *compiler* and is itself a very complex program of computer instructions.

Figure 1.1 shows, in simplified diagrammatic form, the processing of a program through a computing system. The source program is entered, usually

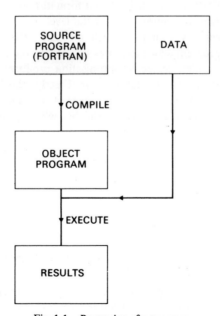

Fig. 1.1   Processing of a program

followed immediately by the data on which the program is to operate. When the computer is ready for a new job, the source program is compiled into the object program, which is then stored and used to control the computations. It is only at this stage, known as *execution*, that the input data are used and the results generated. This broad outline of program processing should be understood, if only to appreciate how various kinds of programming errors can arise, as explained in Chapter 2. It also helps to explain the distinction between executable and non-executable FORTRAN statements. The former define the operations to be performed at the execution stage, whereas the latter provide essential information to the compiler. Examples of non-executable statements include DIMENSION, which indicates how many words of fast store should be allocated to each subscripted variable, and type statements such as REAL and INTEGER, which indicate explicitly how the bits in words associated with particular variables should be interpreted.

At the beginning of every program, before the FORTRAN statements, certain information must be provided to the computer concerning the job it is about to process. The details and format of this job control information depend on the particular computer, but generally the maximum computing time, lines of output and fast core storage must be specified, together with some user identification for accounting purposes. While the cost of computing is generally based on the actual time and quantity of output, together with the requested core storage, it is in the user's interests to provide accurate job control information to expedite the processing of his program.

Among other options in the job control information there is usually a choice of compilers, the main difference being in the level of optimisation applied to the translation process. A compiler which provides little or no optimisation produces an object program which executes slowly, but the compilation is rapid. On the other hand, a high degree of optimisation produces an efficient object program at the expense of compilation time. Clearly, the latter type is most suitable for a well-tested program which is expected to require a lot of execution time. Also, there is usually a compiler available which provides very detailed diagnostic information about the source program and any errors detected during compilation or execution. While such a compiler is very useful for testing new programs, it has the disadvantage of occupying a relatively large amount of extra core store. Program testing is discussed in Chapter 2.

1.4   The FORTRAN Language

FORTRAN, whose name is derived from FORmula TRANslation, is a high-level programming language and is particularly suitable for engineering and scientific applications. FORTRAN coding is very similar to the language of mathematics and is easy to learn, at least in comparison to low-level machine languages. While various levels of FORTRAN have been developed, the version used in this book is FORTRAN IV as described by, for example, McCracken (1972). Higher levels with additional features are available on some computers.

A FORTRAN program consists of a series of statements which are of various types. Arithmetic statements define the numerical operations to be performed and have the appearance of equations, with a single variable on the left hand side of the equals sign. The precise meaning of such a statement should be clearly understood: the arithmetic expression on the right hand side, which may include variables, constants and arithmetic operation symbols, is first evaluated and the result assigned to the storage location of the variable on the left hand side. In this context, variables refer to particular words in the computer core store, and in the FORTRAN coding are represented by alphanumeric names chosen by the programmer. A second type of statement is the input or output variety, which causes either the reading of data from an input magnetic tape, or the writing of results onto the output tape. The input tape is supplied by the card or paper tape reader, and the output goes to the line printer. Arithmetic, input and output statements are executed in the order in which they are arranged, any change in this order requiring control type statements. Such statements can be used to create repeated loops or to transfer execution, either conditionally or unconditionally, to other parts of the program. The foregoing are all executable statements. As discussed in the last section, additional non-executable statements are required to provide information to the compiler, and as a rule should be placed before any executable statements. FORMAT statements, which specify how input and output data are arranged, provide an exception to this rule and may be placed anywhere in the program.

Although FORTRAN is largely independent of the computer used, there are a few features which are machine dependent. For example, the maximum number of subscripts to a variable is not fixed, although it is usually at least three. Also, input READ statements and output WRITE statements refer to particular magnetic tape numbers. All examples in this book, which involve only single input and output tapes, use the numbers 5 and 6 for input and output respectively. Usually it is possible, and indeed necessary, to specify in the job control information which numbers are referenced by the program. The word length of the computer core store also influences programming. Apart from the potential roundoff errors discussed earlier, it is clearly misleading to use output FORMATS giving excessive numbers of significant figures. All the programs in this book were run on a computer with a sixty bit word length, giving at least twelve decimal digit precision, and allowing the storage of up to ten characters per word.

While various input and output devices are available with most computers, attention will be confined here to the input of programs and data on punched cards, one card per statement or line of data, and the output of results on a line printer. Although cards will normally accommodate up to 80 characters, each represented by a column of punched holes, FORTRAN statements can use only the first 72 columns. For reasons of space, all lines of input data and output results are also limited to 72 characters in this book. In general, however, full use should be made of the available widths, which are usually the full 80 characters on data cards, and of the order of 132 characters on line printers. Before punching the cards for a program, the FORTRAN statements should be written out on coding forms having the same format. Although columns 73 to 80 are not available for coding, they may be used for identification purposes such as statement sequence numbering.

In this book no attempt is made to describe the FORTRAN language in detail, which is the province of programming manuals. It is assumed that the reader has some prior knowledge and experience of the language, and access to a FORTRAN manual relevant to his particular computer. The emphasis here is on the efficient use of FORTRAN programming, in conjunction with numerical analysis, for the solution of engineering problems. Nevertheless, the material is arranged so that the first applications in Chapter 3 use only elementary FORTRAN and avoid, for example, subprograms. Meanwhile, Chapter 2 reviews the general approach to programming engineering problems.

2 General Approach to Programming Engineering Problems

Before considering ways of programming particular engineering problems, it is useful to review the general approach to such programming tasks. In this chapter, the main steps in the process are identified and discussed, with particular emphasis on good programming practice and the importance of thorough program testing.

2.1 Steps in a Typical Problem Solution

The main steps in the solution of a typical engineering problem with the aid of a digital computer may be identified as follows.

(1) Define the problem to be solved

As with any engineering analysis, a careful definition of the problem is essential. This often involves the exercise of engineering judgement and experience. For example, if a component or structure is to be designed, this can only be done if the service conditions and design criteria are specified. It is often relevant to consider at a very early stage whether the problem is worth solving on a computer: this question is examined in the next section.

(2) Construct an appropriate mathematical model

An appeal must be made to one or more engineering disciplines to provide the relevant mathematical equations to describe the problem. Simplification of this model may be necessary to make a computer analysis economically justifiable.

(3) Manipulate the equations for numerical solution

The choice of numerical method should be influenced by its suitability for machine computations. A method which is logical and methodical, and therefore easy to program, is to be preferred, even though a large amount of arithmetic may be involved. Hence the extensive use of iterative techniques.

(4) *Construct an algorithm for the solution of the problem*

This algorithm defines the instructions to be executed by the computer, to either produce a solution or decide that none exists. While the number of instructions is finite, many of them may be executed repeatedly, in which case provision is usually made to stop execution after a finite number of cycles. Algorithms are independent of the programming language, but define the structure of the program, from which the coding may be written. Flow charts provide a convenient diagrammatic means for constructing algorithms and are discussed in Section 2.4.

(5) *Choose a computer and programming language*

If a choice of machine is available, the main criteria are size and speed in relation to the complexity of the problem. While the consideration of languages other than FORTRAN is beyond the scope of this book, the reader should be aware that such languages do have their own particular areas of application.

(6) *Check for library programs or subprograms*

Parts of the algorithm are likely to involve standard procedures, for which programs may exist in the computer library. There is a tendency, however, for such programs to be treated as 'black boxes' without the user understanding precisely what they do. Since this book is intended for those learning to program engineering problems, the existence of library programs is largely ignored.

(7) *Divide the algorithm into sections to form subprograms*

These are discussed in Section 2.3.

(8) *Write the coding and punch the cards*

(9) *Build up a list of variable names used in the coding, together with their definitions*

Such program documentation is discussed in Section 2.6.3.

(10) *Test the subprograms*

(11) Test the complete program

The very important subject of program testing is discussed in Section 2.5.

(12) Perform the required runs and interpret the results

With problems of a design type, the computed results usually do not provide one specific answer. Instead, the computer is used to explore a range of possible solutions. It is then up to the user to interpret these in the light of the design criteria.

(13) Write a user's manual

If the program is to be of any lasting value, a detailed description of how it operates and how to use it must be prepared.

To summarise, a computer is only capable of following a carefully defined program of instructions. It is the responsibility of the programmer to ensure that the results obtained lead to a correct solution of his problem. In order to use a computer successfully, all stages of the work must be carefully planned, and the user must have a very sound understanding of the problem, the mathematical model, and the relevant numerical analysis.

2.2   General Economics of Computing

While computers perform arithmetic very much more cheaply than human labour, there are a number of important economic aspects to engineering computing. Because of the enormously high speeds achieved, it is now possible to solve problems that were previously regarded as intractable, since they would take many years to solve by hand. Consequently, ever more complex problems and methods of solution are being examined, until the cost of obtaining solutions becomes excessive. Clearly, the economic advantages of computed solutions should be considered in relation to the costs of developing and using the programs. In developing a program, and particularly one which uses substantial amounts of computing time, there is a tendency to concentrate on reducing this time to a minimum. Naturally, this is highly desirable if the program is to have a long useful life. On the other hand, if it is to be used only a few times the extra cost of refinement, in terms of programming effort and machine time for test runs, may far outweigh the savings.

The cost of computing is not based solely on the length of time required for compilation and execution, but rather on a combination of time, amount of fast core store requested, and the quantity and type of input and output involved. The balance between execution time and core store is important. Methods of solution that are fast to compute often require relatively large amounts of core store, and vice versa.

In many cases, the computer hardware has an important influence on engineering programs. For example, the finite size of the fast store may limit the size or complexity of the problems that can be conveniently handled. At the very least it may impose a particular method of solution. Similarly, if there is a choice of machines, their relative sizes and speeds are important considerations.

One important question to be asked of any piece of engineering computing is whether it is worth doing at all. The original problem may not be sufficiently well-defined to justify a detailed analysis. On the other hand, there is a tendency to develop simplified mathematical models which have only a superficial relevance to the practical problem. It may be cheaper to use another method of resolving the problem, such as trial and error.

2.3 The Use of Subprograms

For large programs the use of subprograms, which in FORTRAN terminology include both SUBROUTINES and FUNCTIONS, offers a number of advantages. Subprograms are independent units, the only communication of information being through their arguments or through COMMON blocks of storage locations. Because of this independence, each subprogram can be developed and tested separately. As a rough guide to length, subprograms containing no more than about fifty statements are usually appropriate. This provides a convenient size of program unit for coding and testing.

In many programs it is necessary to carry out the same series of arithmetic operations, using different variables, at a number of points in the computation. For example, several different numerical integrations using Simpson's rule may be needed. Such operations can be coded as a single subprogram which is called into use when required. In principle, subprograms developed by one programmer can be readily used by another provided that the coding is accompanied by an adequate manual. This is particularly true of library subprograms, when the original programmer is not available for consultation.

2.4 Flow Charts

Before writing the FORTRAN coding, it is necessary to construct some form of flow chart to define the algorithm for the solution of the problem. For experienced programmers the choice between drawn and mental flow charts is a matter of personal taste, but the beginner will often find it helpful to draw such a chart. A typical flow chart is presented for the first case study in the next chapter, and the symbols used are shown in Fig. 2.1. Space considerations forbid the presentation of flow charts for all the examples, but it would be a useful exercise for the reader to construct his own.

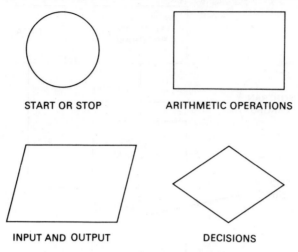

Fig. 2.1   Flow chart symbols

2.5   Program Testing

Program testing is a vitally important stage in the solution of any problem. It is also the most tedious and time consuming, and therefore often neglected. Two general principles should be constantly borne in mind when testing computer programs. Firstly, the program is wrong until proved right under all possible circumstances. Secondly, any errors produced are almost invariably due to faults in the program rather than in the computer. Machine failures, such as the loss of the output from the program, are usually recognisable as such, and the chances of an arithmetic error being made can be ignored.

Programming errors are of three main types, namely language errors, execution errors and algorithm errors. Their effects are detected at different stages in the processing of the program as shown in Fig. 2.2, which is based on Fig. 1.1. Errors in the input data can also cause trouble.

2.5.1 FORTRAN errors   FORTRAN errors are detected during compilation of the object program, and are usually identified by diagnostic messages printed on the listing of the program. Common mistakes include the use of illegal forms of statements and variables, mixing REAL and INTEGER quantities in arithmetic expressions, missing or duplicated statement numbers, and failure to include subscripted variables in a DIMENSION statement. Most FORTRAN errors prevent execution of the program.

Compilers vary considerably in the amount of information they provide about FORTRAN errors and about the program in general. As indicated in Section 1.3, they are designed for different purposes. For testing programs, a

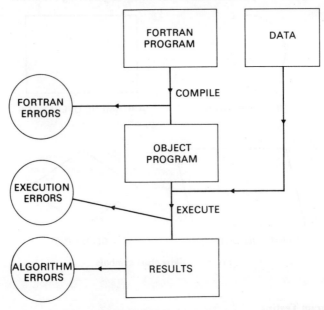

Fig. 2.2　Detection of programming errors during the processing of a program

compiler giving the maximum amount of diagnostic information is to be preferred. Some compilers will even provide comments on the FORTRAN style and suggest ways to improve the efficiency of the program.

2.5.2 Execution errors　Errors which occur during the execution of a program are usually caused by incorrect or incomplete algorithms, which either produce unexpected numbers during the computation or fail to allow for all possible input data. Arithmetic execution errors include the following

(1) Infinite result, when division by zero is attempted.
(2) Complex result, when the raising of a negative number to a REAL exponent is attempted. Complex arithmetic can be handled in FORTRAN, but not by this method.
(3) Overflow or underflow, as described in Section 1.2.1.

Such errors normally cause execution to be terminated and an explanation to be printed, including the location of the relevant statement. With some computers, however, certain errors are bypassed, either with or without a printed warning. For example, the results of underflow or division by zero might be set to zero.

Another common execution error is that of generating subscripts for variables that are either negative or too large. While the maximum array sizes of subscripted variables are specified in DIMENSION statements at the beginning of each program or subprogram, there is often no warning if the subscripts exceed these limits. Since adjacent areas of the core store are thereby unintentionally corrupted, the effects on the program and results are unpredictable, and frequently subtle. A further type of error which may cause trouble during execution, is the use of a variable in an arithmetic expression before a numerical value has been assigned to it.

2.5.3 Algorithm errors All too frequently, faults in the logic of a program cause incorrect results to be produced. Such algorithm errors are often very elusive and demand very careful program testing. They are frequently called 'bugs', possibly to imply that the programmer is less to blame than the computer, and the process of eliminating them is termed *debugging*.

If the results are obviously incorrect, then the logic of both the flow chart and the coding must be checked. It is usually very helpful to print out intermediate results to trace the execution of the program. The extra output statements can be eliminated after testing. If the results from a few sets of input data look reasonable, it is very tempting to accept the program as being correct. It is essential, however, to devise a suitable series of test cases for which the results are known, if necessary by working through the calculation manually. A wide range of input data should be tested, and should be chosen to ensure that every part of the program is executed at least once.

Apart from algorithm errors, the computed results may not provide solutions to the original problem because of faults in either the mathematical model or the numerical analysis. For example, an iterative numerical method of solution may not converge as the number of cycles is increased.

2.5.4 Input data errors The use of invalid input data can often cause execution errors or apparent algorithm errors. Such data are frequently due to mispunched data cards. For example, the INTEGER constant 6 punched one column to the left of its correct position would be read as 60, even though the zero is not punched. This might cause a subscripted variable to exceed its limits as defined in a DIMENSION statement, with unexpected results. Ideally, the program should be designed to neglect all such invalid data by printing a warning and terminating execution. Failing this it is advisable to print out all the data, preferably immediately after they are read in.

2.6   Guidelines for Good Programming

A bad program is one that does not give the correct results, irrespective of how elegant and sophisticated it is. Among possible programs which give the correct results, however, there is a wide range of quality in terms of efficiency, generality and adaptability, and ease of understanding. The main aspects of good programming practice can be considered under the headings of coding style, algorithm efficiency and program documentation.

2.6.1 FORTRAN style

FORTRAN programs should be written in such a way as to make the coding simple to follow and check, and at the same time efficient in terms of execution time and core storage. Some general rules are as follows

(1) Variable names should be readily identifiable with the physical or mathematical quantities they represent.

(2) Arithmetic expressions should not be allowed to become so complicated that many levels of parentheses are required. Such statements should be rewritten as two or more simpler ones.

(3) While division is a somewhat slower arithmetic operation than either addition, subtraction or multiplication, exponentiation with a REAL exponent takes at least ten times longer than the others. Thus, for example, $Y*0.5$ is preferable to $Y/2.$, while $Z**2.$ should be avoided in favour of $Z**2$ or $Z*Z$.

(4) Testing for the exact equality of REAL numbers should be avoided, for the reasons explained in Section 1.2.2.

(5) The repeated evaluation of identical arithmetic expressions using the same numbers should be avoided. For example, as far as possible expressions involved within program loops should be evaluated before the loops are entered.

(6) As discussed in Section 2.3, large programs should be divided into shorter subprograms. Despite their advantages, the use of subprograms involves somewhat longer execution times, due to the time involved in transferring control of execution.

(7) It is more efficient in terms of computing time to transfer data between subprograms by means of COMMON blocks of storage rather than arguments. The latter do, however, offer greater flexibility.

(8) If subprogram arguments are used repeatedly within the subprogram, then it saves time to first assign their values to local variables, although additional core storage is required.

More detailed discussions of these and other aspects of FORTRAN style are given by Kreitzberg and Schneiderman (1972).

2.6.2 Algorithm efficiency

The efficiency of an algorithm can be assessed in terms of the often conflicting criteria of minimum execution time and minimum number of words of core store. In principle, it is possible to compare alternative algorithms by adding up the times involved in all the individual arithmetic operations, and counting the words required. Most computers also have facilities for recording the execution times of various parts of a program. Generally, it is only worthwhile trying to optimise an algorithm if it is to be executed a great many times during the life of the program. Substantial improvements in problem solving efficiency can sometimes be obtained by trying an alternative numerical method of solution, and hence a different algorithm.

2.6.3 Documentation There are three main kinds of program documentation: internal documentation by means of comment statements in the coding, captions and titles in the printed output, and external documentation in the form of user's manuals and flow charts. As a general rule, more documentation should be provided than seems necessary at the time of writing and developing the program. This facilitates both debugging and later use of the program.

Comment statements should be used liberally in the program, both to explain the coding and to separate successive sets of statements for improved readability. At the beginning of each program or subprogram a suitable title and brief explanation of its function should be provided. Additional information, such as a full list of the variable names and their definitions, can also be supplied in comment statements. Careful attention should be paid to the printed output, which should be clear and concise. All numerical data should be labelled with suitable captions, and the number of significant digits specified in the output FORMATS should reflect the accuracy of the input data and the method of solution employed. As indicated in Section 2.5.4, all input data should be printed out, preferably immediately after being read in. Full use should be made of warning messages, possibly followed by termination of execution, whenever an unexpected or unacceptable condition arises in the course of the computation.

When the program has been written and thoroughly tested, a user's manual should be prepared. This should describe in detail the function of the program, and should provide a full list of variable names used in all the subprograms. Finally, a full explanation of how to run the program, including the input data required and the output produced, should be provided.

3 Series, Curve Fitting, Interpolation and Roots of Equations

In this chapter a selection of relatively simple examples of the application of digital computing to the solution of engineering problems is considered. Detailed case studies are described and both the programs and typical results are presented. Only elementary FORTRAN is employed, thus avoiding the use of subprograms and other sophisticated features of the language, which are not introduced until the next chapter.

3.1 Infinite Series

The mathematical solutions to some engineering problems take the form of infinite series. For present purposes it is assumed that such series are convergent, in the sense that not only do the terms become progressively smaller in magnitude, but that their sum tends to a constant value as the number of terms is increased indefinitely. The problem is to obtain a sufficiently accurate numerical approximation to the sum, using only a finite number of terms. This concept of numerical approximation is fundamental to many types of engineering computation, and the error involved in the approximation is called the truncation error. Consider the series

$$S_\infty = \sum_{i=1}^{\infty} a_i \tag{3.1}$$

If the summation is performed over n terms only, then the (truncation) error e and relative error e_r are defined as

$$e = \text{true value} - \text{approximate value} = S_\infty - S_n \tag{3.2}$$

$$e_r = \frac{\text{error}}{\text{true value}} = \frac{S_\infty - S_n}{S_\infty} \tag{3.3}$$

18

If S_∞ is not known there is no direct way of determining either error, and the following approximations are made

$$e \approx a_{n+1}, \qquad e_r \approx \frac{a_{n+1}}{S_n} \tag{3.4}$$

Convergence is assumed when $|e_r| < \alpha$, where α is some suitably small tolerance. In general this tolerance must be made substantially smaller than the actual relative error required.

3.2   Case Study: Shape Factor for Flow Along a Rectangular Channel

Figure 3.1 shows the geometry and cartesian coordinates for the laminar flow of a viscous newtonian fluid along a rectangular channel. The velocity of flow, w, is in the z-direction normal to the cross-section. In practical applications, a

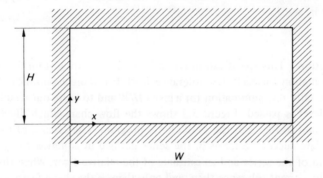

Fig. 3.1   Rectangular channel geometry and coordinates

relationship between volumetric flow rate and pressure drop along the channel is usually required. It has been shown (Fenner, 1970) that for steady flow in a uniform channel the downstream pressure gradient is constant over the cross-section, and is given by

$$\frac{\partial p}{\partial z} \equiv P_z = \mu \left(\frac{\partial^2 w}{\partial x^2} + \frac{\partial^2 w}{\partial y^2} \right) \equiv \mu \nabla^2 w \tag{3.5}$$

where μ is the viscosity of the fluid. This type of mathematical equation, which arises in many engineering problems, is considered in more detail in Chapter 6. Given the condition $w = 0$ on the flow boundaries, a series solution may be obtained for w, from which the volumetric flow rate Q is found as

$$Q = -\frac{P_z W H^3 F}{12\mu} \tag{3.6}$$

where F is the shape factor, given by

$$F = 1 - 192 \frac{H}{W} \sum_{i=1,3,5\ldots}^{\infty} \frac{\tanh\left(\pi i W/2H\right)}{\pi^5 i^5} \qquad (3.7)$$

F is called a shape factor because it depends only on the ratio of channel depth to width, H/W. The calculation of Q using equation (3.6) is simple, so attention is confined to programming the series summation for F.

3.2.1 Problem specification   The numerical values of F are to be obtained for convergence tolerances of 10^{-m}, $m = 1,2,3 \ldots 10$, applied to the series summation. The computer program is to read H/W and the value of the term counter, i, at which the summation is to be terminated if convergence is not achieved. The printed output is to show, for each tolerance level, the current value of i, the term last added, the sum of the truncated series, the approximate relative error, and F.

3.2.2 Solution   This specification could be met by starting from the beginning of the summation for each new tolerance level. It is more economical, however, to perform only one summation for a given H/W and to write out results as each tolerance level is passed. Figure 3.2 shows the flow chart which illustrates the algorithm used in Fig. 3.3, the FORTRAN program. The essence of the process is to calculate successive terms in the series defined in equation (3.7), also the current sum of the series and an estimate of the relative error. When this error is less than the current tolerance then, and only then, is the shape factor computed and written out. Meanwhile, the next smaller tolerance is selected and the summation continued until this too is satisfied. The process is finally terminated when either all the tolerances have been satisfied, or the specified maximum number of terms have been used.

In writing programs for the case studies in this book, the general guidelines of Section 2.6 have been adhered to as far as possible. In order to further improve readability, however, a uniform system of statement numbering is used. Those executable statements requiring numbers are numbered in sequence from 1, while input and output FORMAT statements are numbered from 51 and 61 respectively.

The program variable names used in Fig. 3.3 should be self-explanatory. For example, the array TOLER contains the ten tolerance levels to be examined, and PI is π correct to nine significant figures. HOW (H over W) is used for H/W, TERM for the current term in the series, and SUM for its current sum. TANH is the mathematical function for hyperbolic tangent, and ERROR is the approximate relative error defined in equations (3.4). IT is the tolerance counter

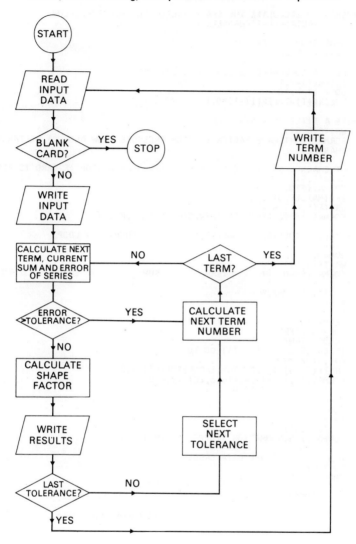

Fig. 3.2   Flow chart for the series summation program

(subscript) and I the term counter i, while PII is the product πi: note the conversion of I to REAL type before the product can be formed.

It is worth noting the different methods used to vary I and IT. The term counter is made the index of a DO loop with the required regular increment of two, whereas IT is increased by IT = IT + 1 only when a new tolerance level is required, which occurs at irregular intervals of I. The summation DO loop illustrates the use of a dummy CONTINUE to allow transfer to the next value of I if the tolerance is not satisfied. Also, if the loop is completed by I reaching

```
C   PROGRAM TO CALCULATE THE SHAPE FACTOR FOR PRESSURE FLOW
C   ALONG A RECTANGULAR CHANNEL.
C
      DIMENSION TOLER(10)
      PI=3.14159265
C
C   STORE THE TOLERANCE LEVELS TO BE EXAMINED.
      TOLER(1)=0.1
      DO 1 IT=2,10
   1  TOLER(IT)=TOLER(IT-1)*0.1
C
C   WRITE A TITLE ON A NEW PAGE.
      WRITE(6,61)
  61  FORMAT(63H1SHAPE FACTOR (F) FOR PRESSURE FLOW ALONG A RECTANGULAR
     1CHANNEL)
C
C   INPUT CHANNEL SHAPE AND MAXIMUM TERM NUMBER - STOP IF CARD IS BLANK.
   2  READ(5,51) HOW,IMAX
  51  FORMAT(F10.0,I5)
      IF(IMAX.LE.0) STOP
      WRITE(6,62) HOW
  62  FORMAT(28HOCHANNEL DEPTH/WIDTH RATIO =,F8.4)
C
C   INITIALISE TOLERANCE COUNTER AND SET UP SUMMATION LOOP.
      IT=1
      SUM=0.
      WRITE(6,63)
  63  FORMAT(68HO   I       TERM        SUM          ERROR        TOLER
     1       F      /)
      DO 3 I=1,IMAX,2
      PII=I
      PII=PII*PI
      TERM=TANH(0.5*PII/HOW)/PII**5
      SUM=SUM+TERM
      ERROR=TERM/SUM
      IF(ERROR.GT.TOLER(IT)) GO TO 3
      F=1.-192.*HOW*SUM
      WRITE(6,64) I,TERM,SUM,ERROR,TOLER(IT),F
  64  FORMAT(1X,I5,E12.4,E16.8,2E11.3,F12.8)
      IF(IT.EQ.10) GO TO 4
      IT=IT+1
   3  CONTINUE
      I=IMAX
   4  WRITE(6,65) I
  65  FORMAT(28HOSUMMATION STOPPED AFTER I =,I5)
C
C   RETURN TO READ A NEW DATA CARD.
      GO TO 2
      END

0.5        1000
```

Fig. 3.3   Program for shape factor series summation

IMAX, then I must be defined (as equal to IMAX) before being re-used, in this case written out. The method of obtaining the sum of the series is a very common programming device whereby SUM is set to zero before the DO loop, and increased by the value of each term as it is calculated. It is important to note that in testing the value of ERROR against the current tolerance it is assumed that ERROR is positive. In this example, both TERM and SUM are always positive so this assumption is justified, but normally ABS(ERROR) should be tested.

Before any data are read, a title is written, to be printed at the top of a new page: statement 61 illustrates the use of the carriage control character for this purpose, also the use of a continuation card. After reading the data at statement 2, HOW is written out with a caption and double line spacing. A zero or negative value of IMAX causes execution to be terminated. Apart from rejecting invalid data, this also allows deliberate termination by a blank card in the data (IMAX read as zero). With this arrangement an arbitrary number of data cards can be read by statement 2. An alternative method of controlling termination is to first read and store the number of subsequent data cards, but this is generally less convenient. If neither of these methods is adopted the control card following the data (an *end of file* card or equivalent) will be read. Before the summation is started, table headings are witten out for the results which follow: note the slash at the end of FORMAT 63 to give a line space before the table. The required results are written out when each tolerance level is passed. Note that in FORMAT 64 an initial blank field of 1X is included to avoid the possibility of the first digit of a large value of I (10 000 or more) acting as a carriage control character. This precaution should always be taken when printing out numerical fields at the beginning of a line. When the smallest tolerance is satisfied, or when I exceeds IMAX, the last value of I is written out with a caption before a new data card is read and the process is repeated.

3.2.3 Results Figure 3.4 shows typical printed output, for $H/W = 0.5$, obtained from the program of Fig. 3.3, the relevant input data being reproduced below the program. The number of terms of the series required increases considerably as the tolerance is reduced, and it is seen that the sum converges to a constant value. For most practical applications no more than three or four significant figures would be required, so that the series could be truncated at $i = 9$. In this example, the approximate relative errors given by ERROR are in good agreement with the actual relative truncation errors of SUM, found with the aid of its final converged value. This is because the rate of convergence is

SHAPE FACTOR (F) FOR PRESSURE FLOW ALONG A RECTANGULAR CHANNEL

CHANNEL DEPTH/WIDTH RATIO = 0.5000

I	TERM	SUM	ERROR	TOLER	F
3	0.1345E-04	0.3269 0293E-02	0.411E-02	0.100E+00	0.68617319
5	0.1046E-05	0.3270 0750E-02	0.320E-03	0.100E-01	0.68607280
7	0.1944E-06	0.3270 2694E-02	0.595E-04	0.100E-02	0.68605414
9	0.5534E-07	0.3270 3247E-02	0.169E-04	1.000E-04	0.68604883
11	0.2029E-07	0.3270 3450E-02	0.620E-05	1.000E-05	0.68604688
17	0.2301E-08	0.3270 3604E-02	0.704E-06	1.000E-06	0.68604540
27	0.2277E-09	0.3270 3636E-02	0.696E-07	1.000E-07	0.68604509
41	0.2821E-10	0.3270 3641E-02	0.862E-08	1.000E-08	0.68604504
65	0.2816E-11	0.3270 3643E-02	0.861E-09	1.000E-09	0.68604503
101	0.3109E-12	0.3270 3643E-02	0.951E-10	1.000E-10	0.68604503

SUMMATION STOPPED AFTER I = 101

Fig. 3.4   Results from shape factor program

relatively fast, since for large i the corresponding term is nearly proportional to i^{-5}. With slower convergence a tolerance of, say, 10^{-6} might give the sum correct to no more than four significant figures.

Having obtained the result $F = 0.6860$ for a rectangular channel whose depth is half its width, it is necessary to check whether this is correct. In this case the answer can be verified by reference to established results, but at least one result should be checked by hand or desk machine calculation. A further test is possible since, in the original physical flow problem, as H/W approaches zero the value of F should approach one (Q is given by equation (3.6) with $F = 1$ if the channel is infinitely wide compared to its depth). On running the program with a very small value of H/W, this result is indeed obtained, provided an accurate value of π is used. Before the program can be said to be fully tested and ready for use with any input data it must be tested over the full practical range. This range is $\delta \leqslant H/W \leqslant 1$, where $\delta \to 0$, for if $H > W$ then H and W may be interchanged in equations (3.6) and (3.7) (for example, the present result could be applied to a channel with $H/W = 2$). It should be noted that $H/W = 0$ cannot be used as this would involve a division by zero in the evaluation of TERM. The program could be altered to cater for this possibility by setting $F = 1.0$ and avoiding the series summation.

In printing out SUM and F to eight significant figures and discussing only truncation errors, it is assumed that these are the only appreciable errors. The possibility of roundoff errors, which are discussed in Section 1.2.2, should also be considered. Some types of calculation allow accumulated roundoff errors to significantly affect the results if the number of arithmetic operations is large. In the case of series summations such errors may be substantial if the rate of convergence is slow, particularly if individual terms in the series are large compared to its sum. McCracken and Dorn (1964) have demonstrated this effect with the series for evaluating the sine of a large angle. In the present example, however, the rate of convergence is rapid and the computer used retains at least twelve decimal digits. Thus, as in most engineering computations, truncation errors are much more important than roundoff errors.

3.3   Curve Fitting

A common way of representing a set of experimental data is by means of a suitable mathematical equation relating the relevant variables. On a graph of the data points this equation is that of the 'best curve' drawn through the band of points. A straight line is often appropriate for this purpose. The question of whether curve fitting should be used is discussed in Section 3.6.4.

Curve fitting involves two distinct steps: the selection of a suitable form of curve, and the choice of equation parameters to best fit the data. Statistical techniques are used both in the fitting, and in testing whether the chosen form of curve is appropriate, as described by, for example, Hoel (1964). Attention is

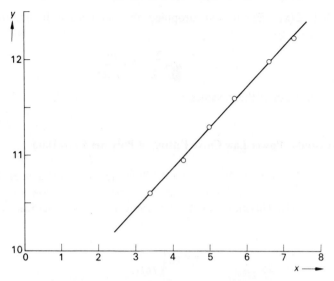

Fig. 3.5   A set of data points and the fitted curve

confined here to fitting a straight line by the well-known method of least squares.

Figure 3.5 shows a set of data points (x_i, y_i), $i = 1, 2, \ldots m$, and the fitted straight line. It is convenient to express the equation of this line as

$$y = b(x - \bar{x}) + c \tag{3.8}$$

where $\bar{x}$ is the mean of the x_i values, b is the slope of the line, and c is the y intercept on $x = \bar{x}$. The values of the parameters b and c are to be chosen such that the sum of the squares of the deviations of the points from the line is minimised. Only the deviations of the y_i values for fixed x_i values will be considered here. In practice, both sets of variables may be subject to experimental scatter. The required sum of squares is

$$S = \sum_{i=1}^{m} (y_i - b(x_i - \bar{x}) - c)^2 \tag{3.9}$$

and the conditions for this to be a minimum are

$$\frac{\partial S}{\partial b} = 0 = - \sum_{i=1}^{m} 2(y_i - b(x_i - \bar{x}) - c)(x_i - \bar{x}) \tag{3.10}$$

$$\frac{\partial S}{\partial c} = 0 = - \sum_{i=1}^{m} 2(y_i - b(x_i - \bar{x}) - c) \tag{3.11}$$

Noting that $\Sigma(x_i - \bar{x}) = 0$ and dropping the summation limits, these two equations reduce to

$$c = \bar{y}, \qquad b = \frac{\Sigma x_i y_i - m\bar{x}\bar{y}}{\Sigma x_i^2 - m\bar{x}^2} \qquad (3.12)$$

where $\bar{y}$ is the mean of the y_i values.

3.4 Case Study: Power-Law Curve Fitting of Polymer Flow Data

As described by, for example, Fenner (1970) the relationship between the shear stress τ and shear rate γ for the laminar non-newtonian flow of a molten polymeric material (plastic) may be represented by the empirical power-law equation

$$\tau = \tau_0 \left(\frac{\gamma}{\gamma_0} \right)^n \qquad (3.13)$$

τ_0 is the effective shear stress at the reference shear rate γ_0, and n is the power-law index. γ_0 can be conveniently chosen as 1 s^{-1}: a form such as equation (3.13) is preferable to $\tau = k\gamma^n$, in which k has units which depend on n.

By taking logarithms, equation (3.13) can be reduced to linear form

$$\ln \tau = n \ln \left(\frac{\gamma}{\gamma_0} \right) + \ln \tau_0 \qquad (3.14)$$

which is identical to equation (3.8) if $y \equiv \ln \tau, x \equiv \ln \gamma, \gamma_0 = 1, b \equiv n$ and $c \equiv \ln \tau_0 + n \bar{x}$.

3.4.1 Problem specification The values of power-law index and effective shear stress at unit shear rate are to be determined by a least-squares curve fitting of six experimental data points, for a polyethylene at 200°C. The program is to be capable of handling up to twenty such points.

3.4.2 Solution Figure 3.6 shows a FORTRAN program for curve fitting. The variable names correspond closely to the symbols used in the above analysis. For example, the arrays GAMMA and TAU store the supplied values of γ and τ respectively, while X and Y store x_i and y_i respectively. M is the integer total number of data points m, RM its REAL counterpart, and I the point counter i. XMEAN, YMEAN, SUMXX and SUMXY are used to accumulate the values of $\bar{x}$, $\bar{y}$, Σx_i^2 and $\Sigma x_i y_i$ respectively, while RN is the power-law index and TAUO is τ_0. ALOG and EXP are the natural logarithm and exponential mathematical functions respectively.

```
C   PROGRAM TO FIT A POWER-LAW CURVE TO POLYMER FLOW DATA BY
C   THE METHOD OF LEAST SQUARES.
C
        DIMENSION GAMMA(20),TAU(20),X(20),Y(20)
        WRITE(6,61)
61      FORMAT(42H1POWER-LAW CURVE FIT FOR POLYMER FLOW DATA/)
C
C   READ THE NUMBER OF DATA POINTS FOLLOWED BY THE SHEAR RATES
C   AND SHEAR STRESSES FOR EACH POINT.
        READ(5,51) M
51      FORMAT(I5)
        IF(M.GE.1.AND.M.LE.20) GO TO 1
        WRITE(6,62) M
62      FORMAT(11H STOP - M =,I5,19H EXCEEDS DIMENSIONS)
        STOP
1       READ(5,52) (GAMMA(I),TAU(I),I=1,M)
52      FORMAT(2F10.0)
        WRITE(6,63) (GAMMA(I),I=1,M)
63      FORMAT(14H SHEAR RATES =,4X,10F9.1)
        WRITE(6,64) (TAU(I),I=1,M)
64      FORMAT(17H SHEAR STRESSES =,10F9.0)
C
C   CALCULATE THE LOGARITHMS OF THE SHEAR RATES AND STRESSES
C   AND STORE IN ARRAYS X AND Y.
        DO 3 I=1,M
        IF(GAMMA(I).GT.0..AND.TAU(I).GT.0.) GO TO 2
        WRITE(6,65)
65      FORMAT(21HOSTOP - NEGATIVE DATA)
        STOP
2       X(I)=ALOG(GAMMA(I))
3       Y(I)=ALOG(TAU(I))
C
C   CALCULATE THE MEANS OF X AND Y AND THE SUMMATIONS OF X**2
C   AND X*Y, HENCE THE POWER-LAW INDEX AND REFERENCE SHEAR STRESS.
        XMEAN=0.
        YMEAN=0.
        SUMXX=0.
        SUMXY=0.
        DO 4 I=1,M
        XMEAN=XMEAN+X(I)
        YMEAN=YMEAN+Y(I)
        SUMXX=SUMXX+X(I)*X(I)
4       SUMXY=SUMXY+X(I)*Y(I)
        RM=M
        XMEAN=XMEAN/RM
        YMEAN=YMEAN/RM
        RN=(SUMXY-RM*XMEAN*YMEAN)/(SUMXX-RM*XMEAN*XMEAN)
        TAUO=EXP(YMEAN-RN*XMEAN)
        WRITE(6,66) RN,TAUO
66      FORMAT(18HOPOWER-LAW INDEX =,F6.3/
1             44H EFFECTIVE SHEAR STRESS AT UNIT SHEAR RATE =,E12.3)
        STOP
        END

     6
  29.1      40400.
  72.9      56900.
  146.      80800.
  291.     109000.
  729.     161000.
  1460.    206000.
```

Fig. 3.6   Program for power-law curve fitting

The value of M is read from the first data card and tested for its validity as a subscript of the arrays. If it is invalid an appropriate warning message is written out, and execution is terminated. There are two reasons for this precaution: the data card might be mispunched, or the program might be run with more data than is allowed by the DIMENSION statement. The effects of exceeding array dimensions are unpredictable and the error often difficult to trace.

The shear rates and stresses are read, one pair per card, using M as an implied DO loop parameter to control the number of cards read. These values are written out with appropriate labels, before any calculations are attempted. This allows the input data to be checked, particularly in the event of a premature termination of execution. The required X and Y values are calculated, after testing for negative data which are unacceptable to ALOG. Finally, the required summations are performed to find n and τ_0, which are written out with appropriate captions, and execution terminated.

3.4.3 Results   Figure 3.7 shows the output obtained from the program and input data shown in Fig. 3.6. Both the data points ($\ln\gamma$, $\ln\tau$) and fitted curve are plotted in Fig. 3.5. The results for n and τ_0 are printed out to three significant

```
POWER-LAW CURVE FIT FOR POLYMER FLOW DATA

SHEAR RATES =            29.1      72.9      146.0     291.0     729.0    1460.0
SHEAR STRESSES =      40400.    56900.    80800.  109000.  161000.  206000.

POWER-LAW INDEX = 0.425
EFFECTIVE SHEAR STRESS AT UNIT SHEAR RATE =    0.956E+04
```

Fig. 3.7   Results from curve fitting program

figures only, the maximum justified by the input data. While the program appears to be functioning correctly, the results have been checked by desk machine calculations.

In this study no attempt has been made to test whether the power-law is an appropriate choice of curve for fitting the data, although Fig. 3.5 shows good agreement. The problem of testing 'goodness of fit' is discussed by, for example, Hoel (1964).

3.5   Interpolation of Tables

A frequent and often tedious activity in engineering calculations is the use of data tables. This may require simultaneous interpolation in the two directions of a table. Such tables are generally constructed with sufficiently small intervals of the variables to permit linear interpolation. Higher order interpolation is possible, but is difficult to apply in more than one direction, particularly if the intervals are not constant.

If f is a function tabulated as a series of values $f_{i,j}$ corresponding to particular values x_i and y_j of two independent variables x and y, then by linear interpolation

$$f \approx f_{i,j} + \delta_x(f_{i+1,j} - f_{i,j}) + \delta_y(f_{i,j+1} - f_{i,j}) \tag{3.15}$$

where
$$\delta_x = \frac{x - x_i}{x_{i+1} - x_i}, \qquad \delta_y = \frac{y - y_j}{y_{j+1} - y_j} \tag{3.16}$$

and
$$0 \leqslant \delta_x < 1, \qquad 0 \leqslant \delta_y < 1 \tag{3.17}$$

Equation (3.15) gives the value of f corresponding to arbitrary values x and y of the independent variables.

3.6 Case Study: Interpolation of Steam Tables

The special functions available with most computers obviate the need for mathematical tables in a programmed calculation. There remain, however, some problems where information is usually only available in tabular form, a good example being the thermodynamic properties of steam. In this study, the interpolation of specific enthalpy (h) tables for superheated steam is considered. The data are obtained from Haywood (1968), the independent variables being pressure p, and temperature T.

3.6.1 Problem specification   A computer program is to be prepared for linearly interpolating values of specific enthalpy over the pressure and temperature ranges $6 \leqslant p \leqslant 20$ MNm^{-2} and $500 \leqslant T \leqslant 650°$C.

3.6.2 Solution   Figure 3.8 shows a FORTRAN program for this interpolation. The variables P and T are the particular values of pressure and temperature, and H the required specific enthalpy, which correspond to x, y and f respectively in equations (3.15) to (3.17). The arrays PDATA, TDATA and HDATA are used to store the tabular values of pressure, temperature and enthalpy respectively, that is x_i, y_j and $f_{i,j}$ in the analysis. NPRESS and NTEMP store the number of pressures and temperatures, and serve to define the extent of the table. The subscripts I and J locate the current position in the table, and are equivalent to i and j respectively in the analysis. Similarly, DELP and DELT are equivalent to δ_x and δ_y. The variables NPM and NTM are introduced because the parameters of DO loops must be simple variables rather than expressions.

```
C   PROGRAM FOR LINEAR INTERPOLATION OF STEAM TABLES.
C
        DIMENSION HDATA(10,10),PDATA(10),TDATA(10)
        WRITE(6,61)
    61  FORMAT(37H1LINEAR INTERPOLATION OF STEAM TABLES)
C
C   INPUT DATA FROM TABLES, THEN WRITE OUT.
        READ(5,51) NPRESS,NTEMP
    51  FORMAT(2I5)
        IF(NPRESS.GE.1.AND.NPRESS.LE.10.AND.NTEMP.GE.1.AND.NTEMP.LE.10)
     1      GO TO 1
        WRITE(6,62) NPRESS,NTEMP
    62  FORMAT(31HOSTOP - DATA EXCEEDS DIMENSIONS)
        STOP
    1   READ(5,52) PDATA,TDATA
    52  FORMAT(10F7.0)
        READ(5,52) ((HDATA(I,J),I=1,NPRESS),J=1,NTEMP)
        WRITE(6,63) (PDATA(I),I=1,NPRESS)
    63  FORMAT(22HOPRESSURE (MN/M**2) = ,10F8.0)
        DO 2 J=1,NTEMP
    2   WRITE(6,64) TDATA(J),(HDATA(I,J),I=1,NPRESS)
    64  FORMAT(14H T (CELCIUS) =,F6.0,4X,10F8.0)
        WRITE(6,65)
    65  FORMAT(61HO      P         T       I    J    DELP     DELT
     1 H   )
C
C   READ PRESSURE AND TEMPERATURE.
    3   READ(5,52) P,T
        IF(P.LE.0.) STOP
        IF(P.GE.PDATA(1).AND.P.LE.PDATA(NPRESS).AND.T.GE.TDATA(1).AND.
     1    T.LE.TDATA(NTEMP)) GO TO 4
        WRITE(6,66) P,T
    66  FORMAT(1X,F10.3,F10.3,19H OUTSIDE DATA RANGE)
        GO TO 3
C
C   LOCATE POSITION IN TABLE.
    4   NPM=NPRESS-1
        DO 5 I=1,NPM
        DELP=(P-PDATA(I))/(PDATA(I+1)-PDATA(I))
        IF(DELP.GE.0..AND.DELP.LT.1.) GO TO 6
    5   CONTINUE
        I=NPRESS
    6   NTM=NTEMP-1
        DO 7 J=1,NTM
        DELT=(T-TDATA(J))/(TDATA(J+1)-TDATA(J))
        IF(DELT.GE.0..AND.DELT.LT.1.) GO TO 8
    7   CONTINUE
        J=NTEMP
C
C   INTERPOLATE.
    8   H=HDATA(I,J)
        IF(I.NE.NPRESS) H=H+DELP*(HDATA(I+1,J)-HDATA(I,J))
        IF(J.NE.NTEMP)  H=H+DELT*(HDATA(I,J+1)-HDATA(I,J))
        WRITE(6,67) P,T,I,J,DELP,DELT,H
    67  FORMAT(1X,F10.3,F10.1,2I5,2F10.3,F10.1)
        GO TO 3
        END

    5     4
    6.    8.    10.   15.   20.
    500.  550.  600.  650.
    3422. 3399. 3375. 3311. 3241. 3539. 3520. 3500. 3448. 3394.
    3656. 3640. 3623. 3580. 3536. 3774. 3759. 3745. 3708. 3671.
    6.    500.
    20.   650.
    7.31  578.
    19.1  651.
    17.5  621.
```

Fig. 3.8 Program for interpolation of steam tables

The amount of arithmetic in this program is relatively small, most of the statements being concerned with data input, output and testing. Following the output of the program title, NPRESS and NTEMP are read from the first data card, and tested for validity as array subscripts. Then the arrays PDATA and TDATA are read, with one card for each array. This illustrates how implied DO loops can be avoided by handling entire arrays. While only the first five elements of PDATA and first four of TDATA are used in the program, the remainder are read as zeros from blank fields on the cards. In contrast, the input of HDATA uses double implied DO loops, the inner loop for I being executed first with J = 1, then J = 2 and so on: compare the order of the HDATA elements on the data cards with the printed table in Fig. 3.9. All the data are printed out under the appropriate pressure headings, for each of the temperatures in turn, as in the original steam tables.

Following the output of a table heading for the interpolations, the first values of P and T are read from a card. This is the third READ statement to reference the same FORMAT. Although this FORMAT statement, numbered 52, implies ten fields on the card, only the first two are used for P and T. In the earlier input of HDATA with more than ten elements, the same FORMAT statement was re-used until the READ was complete. A zero or negative value of P terminates execution. The values of P and T are first tested against the pressure and temperature ranges of the stored data. If either is outside its range, a warning is written out and a new card read. The required position in the table is determined by first finding I and DELP. The array PDATA is scanned from I = 1 to I = NPM, stopping short of NPRESS which would cause the use of a subscript outside the range of the table, and calculating DELP from equation (3.16) at each step, until the condition of equation (3.17) is satisfied. If this is never satisfied, then the only remaining possibility is P = PDATA(NPRESS) exactly. Therefore, on completing the DO loop, I is set equal to NPRESS and interpolation in the pressure direction avoided after statement 8. The same process is repeated to find J and DELT, and finally the interpolated value of H is found using equation (3.15). For checking purposes, not only the values of P, T and H, but also those of I, J, DELP and DELT are written out before new values of P and T are read.

LINEAR INTERPOLATION OF STEAM TABLES

PRESSURE (MN/M**2) =		6.	8.	10.	15.	20.
T (CELCIUS) =	500.	3422.	3399.	3375.	3311.	3241.
T (CELCIUS) =	550.	3539.	3520.	3500.	3448.	3394.
T (CELCIUS) =	600.	3656.	3640.	3623.	3580.	3536.
T (CELCIUS) =	650.	3774.	3759.	3745.	3708.	3671.

P	T	I	J	DELP	DELT	H
6.000	500.0	1	1	0.000	0.000	3422.0
20.000	650.0	5	4	1.000	1.000	3671.0
7.310	578.0	1	2	0.655	0.560	3592.1
19.100	651.0	OUTSIDE	DATA	RANGE		
17.500	621.0	4	3	0.500	0.420	3611.8

Fig. 3.9   Results from interpolation program

3.6.3 Results Figure 3.9 shows the output obtained from the program and input data shown in Fig. 3.8. Of the five interpolations attempted, three are test cases. The first two test the program's handling of the maximum and minimum pressures and temperatures, and the fourth supplies invalid data. The remaining two can be checked by hand calculation.

While this case study considers the interpolation of only a very small area of one steam table, the method can be extended to much larger tables. A program of this type might form a part of a much bigger program for thermodynamic calculations.

3.6.4 Interpolation versus curve fitting In the last two case studies the interpolation and curve fitting of numerical data have been considered. For engineering applications the question sometimes arises as to which is preferable. For example, there are cases of curve fitting being attempted with equations, such as polynomials and other more complicated mathematical functions, involving ten or more parameters. While this is sometimes justified if integration or differentiation of the data is required, often more accurate results can be obtained more quickly by storing the actual data and interpolating as required. In the example studied in Section 3.4, if the only use of the fitted curve is to enable τ to be found for a given γ, or vice versa, then interpolation between the points would probably be adequate. In practice, however, the power-law index finds other uses as a material parameter.

Curve fitting is useful for characterising relatively dense bands of data points, which may be subject to appreciable scatter, by means of a single relatively simple curve. Interpolation is more suitable for relatively sparsely distributed, but more accurate, points which may follow a complicated curve.

3.7   Roots of Nonlinear Equations

There are many engineering examples of nonlinear equations which cannot be solved analytically. Linear equations in one variable are trivial, and simultaneous linear equations are considered in Chapter 5. This section is concerned with nonlinear equations in one variable, multi-variable examples being outside the scope of this book. The general form of equation to be considered is

$$G(x) = 0 \qquad (3.18)$$

where G is an arbitrary function. There are many methods of solution for x, two of the most important being those of functional iteration and Newton-Raphson. Another method is described in Section 4.6.

3.7.1 The functional iteration method   By appropriate rearrangement, equation (3.18) can be written as

$$x = \phi(x) \qquad (3.19)$$

There are normally several ways of doing this, as demonstrated by the case study in Section 3.8. The functional iteration method uses this equation to make successively better approximations to the required root x_r. Thus

$$x_{n+1} = \phi(x_n) \qquad (3.20)$$

where the subscript n indicates the value of x after the nth iteration. Clearly, some initial estimate x_0 must be supplied. The error and relative error, similar to the definitions in equations (3.2) and (3.3), are

$$e = x_r - x_{n+1}, \qquad e_r = e/x_r \qquad (3.21)$$

Since x_r is not known in advance, the following is used as a test for convergence of the iteration

$$\left| \frac{x_{n+1} - x_n}{x_{n+1}} \right| < \alpha \qquad (3.22)$$

where α is a suitable tolerance. As the magnitude of the actual relative error may be larger than α by an unknown amount, α should be small. If the process converges very slowly, then although the relative change tested in equation (3.22) is small, x_{n+1} may still be arbitrarily far from x_r. In this case an additional test such as $|G(x_{n+1})| < \beta$ should be made, where β is another small tolerance.

In common with most other methods for solving nonlinear equations, there are two further difficulties. The first is that instead of converging to a constant value, x may diverge from the root. Figures 3.10a and 3.10b illustrate the itera-

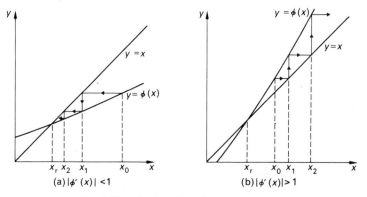

Fig. 3.10   The functional iteration process

tion process for $0 < \phi'(x) < 1$ and $\phi'(x) > 1$ respectively, the first being convergent and the second divergent. Similar diagrams can be drawn for negative $\phi'(x)$, from which it is possible to state the condition for convergence of the iteration as

$$|\phi'(x_r)| < 1 \qquad (3.23)$$

In general, convergence will only be obtained if this condition holds not only at the root itself but also at each iterated value of x, and a poor choice of x_0 may in itself cause divergence. The second difficulty is that the root obtained may be only one of many, and therefore not necessarily the one required. This can often be overcome by a different choice of x_0, although convergence for one root is no guarantee of convergence for any of the others. In some problems, imaginary and complex roots are important, but they are not considered here.

3.7.2 The Newton–Raphson method Starting from equation (3.18), the Newton–Raphson method uses the approximation

$$G(x_r) = 0 \approx G(x_0) + (x_r - x_0)\, G'(x_0) \qquad (3.24)$$

where x_0 is an approximation to the required root x_r. Thus an iteration formula can be constructed

$$\delta x = - \frac{G(x_n)}{G'(x_n)} \qquad (3.25)$$

$$x_{n+1} = x_n + \delta x$$

and a convergence test such as $|\delta x / x_n| < \alpha$ is appropriate. The remarks made above about very slow convergence and the possible need to test $|G(x_{n+1})|$ apply equally here. Figure 3.11 illustrates the progress of a typical Newton–Raphson process.

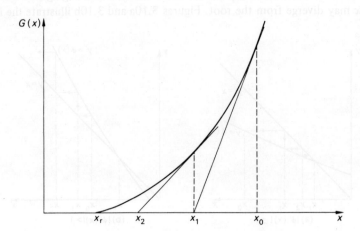

Fig. 3.11   The Newton–Raphson process

3.8 Case Study: Friction Factor for Turbulent Flow in a Smooth Pipe by Functional Iteration Method

The generally accepted empirical equation relating friction factor, f, and Reynolds number, Re, for turbulent flow in a smooth-walled pipe is

$$1/\sqrt{f} = 2{\cdot}0 \log_{10}(Re\sqrt{f}) - 0{\cdot}80 \qquad (3.26)$$

$$f = \frac{\Delta p \ d}{\tfrac{1}{2}\rho \bar{V}^2}, \qquad Re = \frac{\rho \bar{V} d}{\mu}$$

where Δp is the pressure drop per unit length of pipe, d is the internal diameter, $\bar{V}$ the mean velocity, and ρ and μ are the density and viscosity of the fluid respectively. Within the practical ranges $5 \times 10^3 \leqslant Re \leqslant 5 \times 10^6$ and $0{\cdot}008 \leqslant f \leqslant 0{\cdot}04$, there is only one root for f, given the value of Re.

Equation (3.26) can be arranged in functional iteration form as

$$f = \phi(f) = (2{\cdot}0 \log_{10}(Re\sqrt{f}) - 0{\cdot}80)^{-2} \qquad (3.27)$$

The derivative $\phi'(f)$ can be obtained analytically, and the iterative process tested for convergence using equation (3.23). Since the problem is to be programmed, however, the work involved in differentiation is unnecessary as the derivative can be approximated by

$$\phi'(f) \approx \frac{\phi(f + \delta) - \phi(f)}{\delta} \qquad (3.28)$$

where δ is a suitably small number.

3.8.1 Problem specification
A computer program is to be written to calculate friction factors using the functional iteration equation (3.27), for Reynolds numbers of 5×10^3, 5×10^4, 5×10^5, and 5×10^6, and a convergence tolerance of $\alpha = 10^{-6}$. At each iteration a test for possible divergence of the solution is to be made.

3.8.2 Solution
Figure 3.12 shows a FORTRAN program for this iterative method. The variable names include RE and FO for Reynolds number and initial estimate of f respectively, while F is f and F1 and F2 are temporary stores for f (that is, ϕ in equation (3.27)) used in the evaluation of $\phi'(f)$. TOLER is the convergence tolerance α, DPHI is $\phi'(f)$, ITER is the iteration counter, and ERROR is the relative change in f equivalent to the expression in equation (3.22). ALOG10 is the mathematical function for logarithm to the base 10, and IPRINT controls the amount of printed output. IPRINT=0 gives output only after the last iteration, while any other value gives results after every iteration.

```
C    PROGRAM TO CALCULATE THE FRICTION FACTOR FOR TURBULENT FLOW IN A
C    SMOOTH PIPE, GIVEN THE REYNOLDS NUMBER.
C    ***** FUNCTIONAL ITERATION METHOD *****
C
     WRITE(6,61)
 61  FORMAT(52H1FRICTION FACTOR FOR TURBULENT FLOW IN A SMOOTH PIPE//
    1 54H          RE            FO          DPHI     ITER      F     /)
     TOLER=1.E-6
C
C    INPUT REYNOLDS NUMBER, INITIAL ESTIMATE OF F, AND PRINT CONTROL.
 1   READ(5,51) RE,FO,IPRINT
 51  FORMAT(2E15.4,I5)
     IF(RE.LE.0.) STOP
C
C    SET UP THE ITERATION LOOP.
     F=FO
     DO 3 ITER=1,50
     IF(F.LT.0.001) F=0.001
     F1=1./(2.0*ALOG10(RE*SQRT(F))-0.80)**2
C
C    CALCULATE GRADIENT OF ITERATION FUNCTION FOR DIVERGENCE TEST.
     F=F+TOLER
     F2=1./(2.0*ALOG10(RE*SQRT(F))-0.80)**2
     F=F-TOLER
     DPHI=(F2-F1)/TOLER
     IF(ABS(DPHI).LT.1.) GO TO 2
     WRITE(6,62) RE,FO,DPHI,ITER
 62  FORMAT(1X,3E12.4,I5,19H SOLUTION DIVERGING)
     GO TO 1
 2   ERROR=ABS(F1-F)
     IF(ABS(F1).GT.1.E-10) ERROR=ERROR/ABS(F1)
     IF(ERROR.LT.TOLER) GO TO 4
     F=F1
 3   IF(IPRINT.NE.0) WRITE(6,65) DPHI,ITER,F
C
C    NORMAL EXIT FROM ITERATION LOOP INDICATES FAILURE TO CONVERGE.
     ITER=50
     WRITE(6,63) RE,FO,DPHI,ITER
 63  FORMAT(1X,3E12.4,I5,17H - NO CONVERGENCE)
     GO TO 1
C
C    OUTPUT CONVERGED RESULTS.
 4   WRITE(6,64) RE,FO,DPHI,ITER,F
 64  FORMAT(1X,3E12.4,I5,E12.4)
 65  FORMAT(25X,E12.4,I5,E12.4)
     GO TO 1
     END

     5.0000E+03      0.0400E+00
     5.0000E+04      0.0400E+00       1
     5.0000E+05      0.0400E+00
     5.0000E+06      0.0400E+00
```

Fig. 3.12   Program for the functional iteration method

Following the output of the program title and table heading, the first data card containing the values of RE, FO and IPRINT is read. Execution is terminated if RE is zero (blank) or negative. The value of FO is transferred to F so that the former may be retained for writing out, and the iteration loop is set up for a maximum of fifty cycles. It is not necessary to use a subscripted variable for f, as the results of previous iterations are not required and may be over-written. Since very small or negative values of F (outside the practical range) cause execution errors in the iteration function, a minimum value of 0·001 is imposed. F1 is calculated from equation (3.27) as the new value of f.

Then F is increased by an amount equal to TOLER (a suitably small value for δ in equation (3.28)), and a second value F2 obtained before the original value of F is restored. This is merely a convenient way of keeping the expressions for F1 and F2 identical, permitting card duplication. $\phi'(f)$ is obtained and if it fails to satisfy the condition of equation (3.23), the iteration is abandoned. For a converging solution, ERROR is calculated (note the avoidance of division by zero, which although very unlikely here, would cause an execution error) and the iteration stopped if this is less than the convergence tolerance. Otherwise, the new value of F is stored and if necessary the current results are written out. If there is no convergence after fifty cycles and the DO loop is completed, the counter ITER is defined and an appropriate message is written out.

3.8.3 Results   Figure 3.13 shows the output obtained from the program and input data shown in Fig. 3.12. FO is chosen as 0·04, but this choice is not critical. For the four values of Re used, convergence was obtained in seven or

FRICTION FACTOR FOR TURBULENT FLOW IN A SMOOTH PIPE

RE	FO	DPHI	ITER	F
0.5000E+04	0.4000E−01	−0.1680E+00	8	0.3740E−01
		−0.5818E−01	1	0.1929E−01
		−0.1381E+00	2	0.2111E−01
		−0.1241E+00	3	0.2087E−01
		−0.1257E+00	4	0.2090E−01
		−0.1255E+00	5	0.2089E−01
		−0.1256E+00	6	0.2089E−01
		−0.1256E+00	7	0.2089E−01
0.5000E+05	0.4000E−01	−0.1256E+00	8	0.2089E−01
0.5000E+06	0.4000E−01	−0.9964E−01	8	0.1316E−01
0.5000E+07	0.4000E−01	−0.8232E−01	7	0.8982E−02

Fig. 3.13   Results from functional iteration program using equation (3.27)

eight cycles (results are printed after every cycle for $Re = 5 \times 10^4$ only), and in every case $|\phi'(f)|$ is less than one. The calculated friction factors can be checked by substitution into equation (3.26).

In this example, the functional iteration method is well suited to solving the nonlinear equation, but it is worth considering the alternative choice of function. Equation (3.26) could have been rearranged as

$$f = \phi(f) = (Re)^{-2} \, 10^{(0·80 + 1/\sqrt{f})} \qquad (3.29)$$

If appropriate FORTRAN statements for F1 and F2 are used in the program, then the output shown in Fig. 3.14 is obtained, using the same input data. The iteration processes are all divergent, which illustrates the main disadvantage of the method.

FRICTION FACTOR FOR TURBULENT FLOW IN A SMOOTH PIPE

RE	FO	DPHI	ITER	F
0.5000E+04	0.4000E-01	-0.3632E+01	1	SOLUTION DIVERGING
		-0.3632E-01	1	0.2524E-03
0.5000E+05	0.4000E-01	-0.3783E+28	2	SOLUTION DIVERGING
0.5000E+06	0.4000E-01	-0.3783E+26	2	SOLUTION DIVERGING
0.5000E+07	0.4000E-01	-0.3783E+24	2	SOLUTION DIVERGING

Fig. 3.14   Results from functional iteration program using equation (3.29)

3.9   Case Study: Friction Factor for Turbulent Flow in a Smooth Pipe by Newton–Raphson Method

As a comparison with the functional iteration method, the Newton–Raphson method is applied to equation (3.26), and the function G of equation (3.24) is defined as

$$G(f) = 1/\sqrt{f} + 0.80 - 2.0 \log_{10}(Re\sqrt{f}) \tag{3.30}$$

While $G'(f)$ can be obtained analytically, it is again convenient to evaluate it numerically by an approximation of the type given in equation (3.28).

A FORTRAN program for the Newton–Raphson method is shown in Fig. 3.15, and is very similar in terms of general organisation the one shown in Fig. 3.11. The only new variable names are GO and G1 for temporary stores used in the calculation of $G'(f)$, which is stored in DG, and DELTAF is δf (equivalent to δx in equation (3.25)).

3.9.1 Results   Figure 3.16 shows the output from the program and input data shown in Fig. 3.15. The same converged values of f are obtained as in Fig. 3.12. The number of iteration cycles is more dependent on the closeness of the initial estimate FO, but is of the same order as in the functional iteration method.

3.9.2 Comparison of the methods   A thorough comparison of the functional iteration, Newton–Raphson and other methods for solving nonlinear equations is beyond the scope of this book. The reader is referred to numerical analysis texts such as Ralston (1965) or Williams (1972). Nevertheless, some general observations can be made. The functional iteration method may converge in fewer cycles than Newton–Raphson, though not significantly so in the example. It needs less arithmetic as it involves only one function evaluation per cycle (assuming the gradient calculation is omitted), rather than a function and its derivative. This is an important consideration for hand calculations, but generally much less so when using a computer. The disadvantage of the functional iteration method is that it is often difficult or even impossible to find

```
C  PROGRAM TO CALCULATE THE FRICTION FACTOR FOR TURBULENT FLOW IN A
C  SMOOTH PIPE, GIVEN THE REYNOLDS NUMBER.
C  ***** NEWTON-RAPHSON METHOD *****
C
      WRITE(6,61)
61    FORMAT(52H1FRICTION FACTOR FOR TURBULENT FLOW IN A SMOOTH PIPE//
     1 42H          RE          FO      ITER      F      /)
      TOLER=1.E-6
C
C  INPUT REYNOLDS NUMBER, INITIAL ESTIMATE OF F, AND PRINT CONTROL.
1     READ(5,51) RE,FO,IPRINT
51    FORMAT(2E15.4,I5)
      IF(RE.LE.0.) STOP
C
C  SET UP THE ITERATION LOOP.
      F=FO
      DO 2 ITER=1,50
      IF(F.LT.0.001) F=0.001
      G0=1./SQRT(F)+0.80-2.0*ALOG10(RE*SQRT(F))
      F=F+TOLER
      G1=1./SQRT(F)+0.80-2.0*ALOG10(RE*SQRT(F))
      F=F-TOLER
      DG=(G1-G0)/TOLER
      DELTAF=-G0/DG
      ERROR=ABS(DELTAF)
      IF(ABS(F).GT.1.E-10) ERROR=ERROR/ABS(F)
      IF(ERROR.LT.TOLER) GO TO 3
      F=F+DELTAF
2     IF(IPRINT.NE.0) WRITE(6,64) ITER,F
C
C  NORMAL EXIT FROM ITERATION LOOP INDICATES FAILURE TO CONVERGE.
      ITER=50
      WRITE(6,62) RE,FO,ITER
62    FORMAT(1X,2E12.4,I5,17H - NO CONVERGENCE)
      GO TO 1
C
C  OUTPUT CONVERGED RESULTS.
3     WRITE(6,63) RE,FO,ITER,F
63    FORMAT(1X,2E12.4,I5,E12.4)
64    FORMAT(25X,I5,E12.4)
      GO TO 1
      END

      5.0000E+03      0.0400E+00
      5.0000E+04      0.0400E+00      1
      5.0000E+05      0.0400E+00
      5.0000E+06      0.0400E+00
```

Fig. 3.15   Program for the Newton-Raphson method

FRICTION FACTOR FOR TURBULENT FLOW IN A SMOOTH PIPE

RE	FO	ITER	F
0.5000E+04	0.4000E-01	4	0.3740E-01
		1	0.1001E-01
		2	0.1627E-01
		3	0.2012E-01
		4	0.2087E-01
		5	0.2089E-01
0.5000E+05	0.4000E-01	6	0.2089E-01
0.5000E+06	0.4000E-01	9	0.1316E-01
0.5000E+07	0.4000E-01	8	0.8982E-02

Fig. 3.16   Results from Newton-Raphson program

a convergent formulation of the problem. Therefore, unless the amount of extra arithmetic in function evaluation is excessive, the Newton–Raphson method is often to be preferred. While this method is more likely to be convergent there are cases which it does not converge.

Finally, the possibility of roundoff errors affecting the accuracy of the results obtained by either method should be considered. In each iteration, using either equation (3.20) or equation (3.25), the latest estimate, x_{n+1}, of the root depends only on the last value, x_n, as an 'initial' approximation. Therefore, the final roundoff error is merely the error involved in the last iteration, and is consequently negligible. This is generally true of iterative techniques, and is one of their main advantages over other methods.

4 Integration and Ordinary Differential Equations

The mathematical models for many engineering problems are expressed in terms of differential equations. In general, these are partial differential equations involving more than one independent variable, and methods for solving such equations are considered in Chapter 6. In this chapter, attention is confined to ordinary differential equations, which involve only one independent variable.

Ordinary differential equations can contain derivatives of any order, but the simplest are first-order equations, which take the general form

$$\frac{\mathrm{d}y}{\mathrm{d}x} = f(x, y) \tag{4.1}$$

where x is the independent variable and y the dependent variable. Higher-order equations are also considered in this chapter, but for computational purposes can be treated as sets of first-order equations. If the functional form of f involves only x, then the process of solving for y is called *integration*. For the purpose of digital computation the range of x must be specified, and the process is then termed definite integration or quadrature.

Whereas in Chapter 3 only elementary FORTRAN was used, in this chapter a set of generally applicable subprograms is developed. One of the main advantages of subprograms is demonstrated by employing the same subprogram in more than one case study.

4.1 Numerical Integration

If the function f in equation (4.1) depends only on x, but is nevertheless difficult or impossible to integrate analytically, then some form of numerical integration must be used. While many methods have been developed for this purpose (see, for example, Conte (1965); Ralston (1965); or Williams (1972)). the most commonly used are the trapezoidal and Simpson's rules.

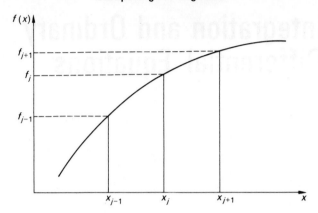

Fig. 4.1   Abscissae and ordinates for numerical integration

4.1.1 Trapezoidal and Simpson's rules   Figure 4.1 shows part of a curve $f(x)$, plotted against x. The definite integral of this function between, say, the abscissae x_j and x_{j+1} can be estimated by approximating the shape of the area under the curve between the corresponding ordinates by a trapezium. Thus

$$y_{j+1} - y_j = \int_{x_j}^{x_{j+1}} f(x)\, dx \approx \tfrac{1}{2}h(f_j + f_{j+1}) \tag{4.2}$$

$$h = x_{j+1} - x_j$$

and the method can be extended to more than one interval of x, as for example

$$y_{j+1} - y_{j-1} \approx \tfrac{1}{2}(x_j - x_{j-1})(f_{j-1} + f_j) + \tfrac{1}{2}(x_{j+1} - x_j)(f_j + f_{j+1}) \tag{4.3}$$

The advantages of this method are its simplicity and the ability to cope with random sizes of interval.

If the x intervals in Fig. 4.1 are equal, then it is preferable to use Simpson's rule

$$y_{j+1} - y_{j-1} = \int_{x_{j-1}}^{x_{j+1}} f(x)\, dx \approx \tfrac{1}{3}h(f_{j-1} + 4f_j + f_{j+1}) \tag{4.4}$$

$$h = (x_j - x_{j-1}) = (x_{j+1} - x_j)$$

which can also be applied to any (even) number of intervals. The main advantage over the trapezoidal rule is that of greater accuracy for the same amount of computation.

4.1.2 Truncation and roundoff errors In the case of the trapezoidal rule, the inaccuracy of equation (4.2) can be visualised as being due to the small area between the straight line joining the points (x_j, f_j) and (x_{j+1}, f_{j+1}), and the actual curve $f(x)$ between these points. This can be described as a truncation error in the sense used in Section 3.1, since the right hand side of equation (4.2) is the first term of an infinite series (derived from a Taylor series as shown, for example, by McCracken and Dorn (1964)). The second term, which provides an estimate of the truncation error, is

$$e_T = -\frac{h^3}{12} f''(x_j) \tag{4.5}$$

Similarly, the second term on the right hand side of equation (4.4), and hence an estimate of the Simpson's rule truncation error, is

$$e_S = -\frac{h^5}{90} f''''(x_j) \tag{4.6}$$

Clearly, for the same (small) h, e_S should be much smaller than e_T. There are functions, however, whose high-order derivatives are large, causing substantial truncation errors. Also, numerical integration is difficult to apply in regions where f is singular and the ordinates tend to infinity, even though the integral is finite.

It is tempting to think that the accuracy of either the trapezoidal or Simpson's rule can be improved indefinitely by increasing the number of intervals in a given range of the independent variable, thus reducing h and the truncation error. In practice, however, maximum accuracies are achieved, beyond which roundoff errors become more significant than truncation errors. Numerical integration involves a number of addition and multiplication operations which is proportional to the number of intervals used. Hence, the total roundoff error, which accumulates from the error involved in each operation as indicated in Section 1.2.2, also increases with the number of intervals. The accuracy of numerical integration techniques is studied, with the aid of a practical example, in Section 4.3.4.

4.2   Case Study: Boundary Layer Data Analysis

When a fluid flows past a solid object there is a region close to the surface, known as a boundary layer, in which the velocity of the fluid is less than that of the free stream. The velocity profile in a boundary layer can be examined experimentally with the aid of a pitot-static tube, which provides a measure of the 'velocity pressure', $\frac{1}{2}\rho u^2$, where u is the local fluid velocity and ρ is its

density. The data obtained may be analysed to find the mass flow rate within
the boundary layer

$$m = \rho u_1 \,\delta \int_0^1 \left(\frac{u}{u_1}\right) \mathrm{d}\left(\frac{y}{\delta}\right) \tag{4.7}$$

where y is the distance from the solid surface, δ is the thickness of the boundary
layer, and u_1 is the free-stream velocity. Also, the momentum deficit of the
boundary layer is given by

$$M = \rho u_1^2 \,\delta \int_0^1 \left(\frac{u}{u_1}\right)\left(1 - \frac{u}{u_1}\right) \mathrm{d}\left(\frac{y}{o}\right) \tag{4.8}$$

4.2.1 Problem specification   A flat plate is held parallel to a fluid stream and a
pitot-static tube is traversed normal to its surface. The following readings of
velocity pressure, in kNm^{-2}, are obtained at ten uniformly spaced intervals

$y/\delta =$	0·1	0·2	0·3	0·4	0·5	0·6	0·7	0·8	0·9	1·0
$\frac{1}{2}\rho u^2 =$	1·29	1·57	1·77	1·92	2·05	2·16	2·26	2·34	2·42	2·50

The density of the fluid is $1 \cdot 19 \text{ kg m}^{-3}$, and the local thickness of the boundary
layer is 12·3 mm. A program is to be written to find the mass flow rate and
momentum deficit in the region traversed. The numerical integration is to be
performed by Simpson's rule, which is to be coded as a subprogram.

4.2.2 Solution   Figure 4.2 shows a main program for this problem, which in
two different statements calls the Simpson's rule subprogram shown in Fig. 4.3.
The subscripted variable VPRESS is used to store the values of velocity pressure,
while FFLOW and FMOMTM store the flow rate and momentum deficit
functions to be integrated according to equations (4.7) and (4.8) respectively.
Note that the dimensioned array size for these three variables is eleven, to allow
for values to be stored for $y = 0$, where the velocity is assumed to be zero. The
variables FLOW and MOMTM are used for the flow rate and momentum deficit
respectively, and since the implicit type of the latter is integer, its name appears
in a REAL statement at the beginning of the program.

In its present form, the program reads and processes only one set of data
before terminating execution. The values of fluid density, RHO, boundary layer
thickness, DELTA, and velocity pressures are read from a single card. Since all
eleven elements of the array VPRESS, including the zero value for VPRESS(1)
at the plate, are read in, only the array name need appear in the READ

```
C   PROGRAM TO ANALYSE EXPERIMENTAL BOUNDARY LAYER DATA.
C
      DIMENSION VPRESS(11),FFLOW(11),FMOMTM(11)
      REAL  MOMTM
      READ(5,51) RHO,DELTA,VPRESS
  51  FORMAT(13F6.0)
      WRITE(6,61) RHO,DELTA,VPRESS
  61  FORMAT(29H1BOUNDARY LAYER DATA ANALYSIS /
     1 10H DENSITY =,F7.3,8H KG/M**3,10X,12H THICKNESS =,F7.2,3H MM /
     2 19H VELOCITY PRESSURES / 11F6.2 )
C
C   EVALUATE VELOCITY RATIOS, FLOW AND MOMENTUM DEFICIT FUNCTIONS.
      U1=SQRT(2.*VPRESS(11)/RHO*1000.)
      DO 1 I=1,11
      UREL=SQRT(VPRESS(I)/VPRESS(11))
      FFLOW(I)=UREL
  1   FMOMTM(I)=UREL*(1.-UREL)
C
C   INTEGRATE FOR MASS FLOW RATE USING SIMPSONS RULE.
      CALL  SIMPV(FFLOW,11,10,0.1,FLOW)
      FLOW=FLOW*RHO*U1*DELTA*0.001
C
C   INTEGRATE FOR MOMENTUM DEFICIT USING SIMPSONS RULE.
      CALL  SIMPV(FMOMTM,11,10,0.1,MOMTM)
      MOMTM=MOMTM*RHO*U1*U1*DELTA*0.001
C
C   OUTPUT THE RESULTS.
      WRITE(6,62) FLOW,MOMTM
  62  FORMAT(17H MASS FLOW RATE =,F7.3,7H KG/M*S /
     1        19H MOMENTUM DEFICIT =,F7.2,8H KG/S**2 )
      STOP
      END
```

Fig. 4.2   Main program for boundary layer data analysis

statement. The input data are immediately written out with appropriate captions and a program title. The free-stream velocity, U1, is calculated from its velocity pressure. The local velocity ratio, UREL (u/u_1), is obtained as the square root of the ratio of the velocity pressures, and is used to define the flow rate and momentum deficit functions. These functions are stored in preparation for numerical integration. The Simpson's rule subprogram is called twice to evaluate the integrals in equations (4.7) and (4.8), from which the required flow rate and momentum deficit are found. The factor 0·001 is used in each case to give the results in units of kg m^{-1}s^{-1} and kg s^{-2} respectively.

4.2.3 The Simpson's rule subprogram   Figure 4.3 shows a SUBROUTINE subprogram, named SIMPV, for Simpson's rule. The names of the variables follow the general notation employed in equation (4.4). The array F stores the values of the function to be integrated over an even number, NSTEP, of steps (intervals) of length H of the independent variable. The final integral is stored in SINT, which is also used during the summation process. This process involves the repeated application of equation (4.4) to successive pairs of steps, the common factor $\frac{1}{3}h$ being applied when the summation is complete. Note that while the number of steps is given by NSTEP, the integration is performed between the ordinates F(1) and F(NSTEP + 1).

This subprogram has been made widely applicable by allowing the array size of F to be specified by the calling program. This can only be done if both F and the dimensioning variable, in this case NDIM, are arguments of the subprogram. In the CALL statements in Fig. 4.2, the arguments corresponding to NDIM define the actual sizes of the supplied arrays FFLOW and FMOMTM. Of the five variables used as arguments of SIMPV, the first four provide the necessary input data and the last one returns the result to the calling program. As there is only one result, a FUNCTION subprogram could have been used instead of the SUBROUTINE type.

```
      SUBROUTINE  SIMPV(F,NDIM,NSTEP,H,SINT)
C
C  SUBROUTINE SUBPROGRAM FOR SIMPSONS RULE - VARIABLE DIMENSIONS.
C
      DIMENSION  F(NDIM)
C
C  CHECK FOR EVEN NUMBER OF STEPS.
      IF(MOD(NSTEP,2).EQ.0) GO TO 1
      WRITE(6,61) NSTEP
   61 FORMAT(45HOEXECUTION TERMINATED IN SIMPV BY ODD NSTEP =,I6)
      STOP
C
C  APPLY SIMPSONS RULE.
    1 SINT=0.
      DO 2 J=2,NSTEP,2
    2 SINT=SINT+F(J-1)+4.*F(J)+F(J+1)
      SINT=SINT*H/3.
      RETURN
      END
```

Fig. 4.3   Subprogram for Simpson's rule

In SIMPV, a preliminary check is made to find whether the supplied number of steps is even. This illustrates both the use of the function MOD for remaindering, and the provision of diagnostic messages in the event of invalid data being encountered. A similar test could be made to ensure that NSTEP is both positive and less than NDIM. The subprogram could be made more general by, for example, allowing integration to start at an ordinate other than F(1); including a single application of the trapezoidal rule if an odd number of steps is encountered; and allowing successive pairs of step lengths to vary. For most applications, however, such refinements are unnecessary.

4.2.4 Results Figure 4.4 shows the results for the present case study, and these may be checked by hand calculation. This particular example of numerical

```
BOUNDARY LAYER DATA ANALYSIS
DENSITY = 1.190 KG/M**3              THICKNESS =  12.30 MM
VELOCITY PRESSURES
  0.00  1.29  1.57  1.77  1.92  2.05  2.16  2.26  2.34  2.42  2.50
MASS FLOW RATE =  0.814 KG/M*S
MOMENTUM DEFICIT =   5.48 KG/S**2
```

Fig. 4.4   Results from boundary layer data analysis program

integration is comparatively simple, but it does serve to illustrate the use of subprograms, as discussed ·in Section 2.3. SIMPV is called from the main program to perform two separate integrations of different functions. In general, the array sizes and step lengths could also be different. This subprogram can now be used in other applications requiring Simpson's rule.

4.3 Case Study: Approximate Buckling Load for a Cantilevered Shaft

When a long thin shaft is subjected to longitudinal compression, it will buckle when the load exceeds a certain critical value. The magnitude of this buckling load depends on the geometry of the shaft, its elastic properties, and the type of constraints applied to it. The present study concerns a cantilevered shaft of circular cross-section, with a compressive load on its free end, as shown in Fig. 4.5. While an accurate solution for the buckling load can be found as described in Section 4.6, an approximate value can be readily obtained using an energy method. This method is discussed in detail by, for example, Timoshenko and Gere (1961).

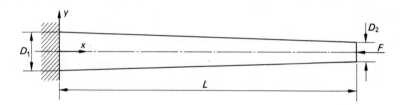

Fig. 4.5   Cantilevered shaft with compressive end load

When the shaft is deflected, energy is stored mainly due to bending, and is given by

$$\text{Energy stored} = \frac{E}{2} \int_0^L I \, (y''(x))^2 \, dx \qquad (4.9)$$

where E is Young's modulus, L is the length of the shaft, and I is the second moment of area of the cross-section about its neutral axis, at a distance x from the fixed end. The deflected shape is given by the function $y(x)$. Similarly, the work done by the end load, F, is

$$\text{Work done} = \frac{F}{2} \int_0^L (y'(x))^2 \, dx \qquad (4.10)$$

As the buckling condition is one of neutral equilibrium, these two quantities are equal, and the critical value of F is

$$F = E \int_0^L I(y'')^2 \, dx \bigg/ \int_0^L (y')^2 \, dx \tag{4.11}$$

Normally the deflected shape is not known, and a suitable form must be assumed. This assumed shape should satisfy the appropriate boundary conditions, but otherwise may take any convenient form, such as a polynomial or a trigonometrical function. The latter type is often preferred because it provides a good approximation to the true shape. Even with a comparatively poor approximation, however, this method provides a good estimate of the buckling load.

4.3.1 Problem specification The magnitude of the end load required to produce buckling of the cantilevered shaft shown in Fig. 4.5 is to be estimated. The shaft diameter decreases uniformly from D_1 at the fixed end to D_2 at the free end. The length of the shaft is $L = 2 \cdot 5$ m, and $D_1 = 0 \cdot 2$ m, while three values of D_2 are to be considered, namely $0 \cdot 2$, $0 \cdot 15$ and $0 \cdot 10$ m. Young's modulus for the material of the shaft is $E = 208$ GN m^{-2}. The following deflected shape is to be assumed

$$y(x) = \delta \, (1 - \cos \, (\pi \, x/2L)) \tag{4.12}$$

where δ is the end deflection. In the case of the parallel shaft, the computed result is to be compared with the exact solution $F = \pi^2 EI/4L^2$.

4.3.2 Solution It should be noted that equation (4.12) satisfies the cantilever constraint conditions $y(0) = y'(0) = 0$. While the deflected shape involves the end deflection, the magnitude of δ is irrelevant: it cancels out in equation (4.11). Buckling occurs when the end load is just sufficient to cause arbitrarily large deflections.

For a circular cross-section, the required second moment of area is

$$I = \pi D^4/64 \tag{4.13}$$

where the diameter D at a distance x along the shaft is given by

$$D = D_1 + x \, (D_2 - D_1)/L \tag{4.14}$$

The required derivatives of y may be obtained from equation (4.12) as

$$y'(x) = \frac{\delta \pi}{2L} \sin \, (\pi \, x/2L) \tag{4.15}$$

$$y''(x) = \frac{\delta \pi^2}{4L^2} \cos \, (\pi \, x/2L) \tag{4.16}$$

Figure 4.6 shows a main program for computing the buckling load according to equation (4.11), using the Simpson's rule subprogram shown in Fig. 4.3. The subscripted variables EF and WF are used to store the energy and work functions to be integrated according to equations (4.9) and (4.10) respectively. The array size for these two variables is set at fifty. This number is stored in the variable NDIM which is used as an argument for the subprogram SIMPV as described in Section 4.2.3. The variables I and L, which are used for the second moment of

```
C   PROGRAM TO ESTIMATE SHAFT BUCKLING FORCE BY ENERGY METHOD.
C
        DIMENSION  EF(50),WF(50)
        REAL  I,L
        NDIM=50
        PI=4.*ATAN(1.)
        WRITE(6,61)
61      FORMAT(43H1     SHAFT BUCKLING FORCE BY ENERGY METHOD //
      1 52H     D1        D2        L          E        NSTEP     F      )
1       READ(5,51) D1,D2,L,E,NSTEP
51      FORMAT(3F10.0,E15.5,I5)
C
C   TEST VALIDITY OF THE NUMBER OF INTEGRATION STEPS.
        IF(NSTEP.GT.0.AND.NSTEP.LT.NDIM) GO TO 2
        STOP
C
C   EVALUATE THE ENERGY AND WORK FUNCTIONS.
2       FACT=0.5*PI/L
        JMAX=NSTEP+1
        DO 3 J=1,JMAX
        X=L*FLOAT(J-1)/FLOAT(NSTEP)
        D=D1+X/L*(D2-D1)
        I=PI*D**4/64.
        DY=FACT*SIN(X*FACT)
        D2Y=FACT*FACT*COS(X*FACT)
        EF(J)=I*D2Y*D2Y
3       WF(J)=DY*DY
C
C   APPLY SIMPSONS RULE TO THE ENERGY AND WORK FUNCTIONS IN TURN.
        H=L/FLOAT(NSTEP)
        CALL  SIMPV(EF,NDIM,NSTEP,H,EINT)
        CALL  SIMPV(WF,NDIM,NSTEP,H,WINT)
        F=E*EINT/WINT
C
C   OUTPUT THE RESULTS.
        WRITE(6,62) D1,D2,L,E,NSTEP,F
62      FORMAT(1X,3F7.2,E12.4,I6,E12.4)
        GO TO 1
        END
```

Fig. 4.6   Main program for the estimation of shaft buckling load

area and shaft length respectively, are declared in a REAL statement. Note the use of the arctangent function to find the value of π, which is stored in the variable PI. Alternatively, the value of π could be specified to an appropriate number of significant figures as in Fig. 3.3.

After writing out an appropriate program title and heading for the table of results, the shaft end diameters D1 and D2, length L, and Young's modulus E are read in, together with NSTEP, the number of steps along the shaft for numerical integration. Unacceptable values of NSTEP cause execution to be terminated, and a blank data card therefore stops the program. The variable FACT is employed to store the frequently used factor $\pi/2L$, to avoid unnecessary repeated calculations within the subsequent DO loop. For each of the NSTEP + 1 positions along the shaft, the diameter and second moment of area are calculated using equations (4.14) and (4.13) respectively. Then $y'(x)$ and $y''(x)$, represented by DY and D2Y, are found using equations (4.15) and (4.16) (taking δ as one). Finally, the values of the energy and work functions are determined.

The step length (x interval) is stored in the variable H and the subprogram SIMPV is called twice to evaluate the two integrals defined in equations (4.9) and (4.10). The results are returned to the main program stored in the variables EINT and WINT. The buckling load, F, is obtained using equation (4.11) and is written out, together with the input data.

4.3.3 Results Figure 4.7 shows the results obtained from the program shown in Fig. 4.6, for the required range of shaft geometry. For the first three results, in which D_2 is varied, 32 integration steps are used. The result for the parallel shaft agrees to four significant figures with the exact solution $F = \pi^2 EI/4L^2 = 0.6449 \times 10^7$ N. This excellent agreement is due to the fact that equation (4.12) for the assumed deflected shape is the true shape for a parallel shaft.

SHAFT BUCKLING FORCE BY ENERGY METHOD

D1	D2	L	E	NSTEP	F
0.20	0.20	2.50	0.2080E+12	32	0.6449E+07
0.20	0.15	2.50	0.2080E+12	32	0.4821E+07
0.20	0.10	2.50	0.2080E+12	32	0.3671E+07
0.20	0.10	2.50	0.2080E+12	16	0.3671E+07
0.20	0.10	2.50	0.2080E+12	8	0.3671E+07
0.20	0.10	2.50	0.2080E+12	4	0.3662E+07
0.20	0.10	2.50	0.2080E+12	2	0.3510E+07

Fig. 4.7   Estimated buckling loads

The last four results in Fig. 4.7 demonstrate the effect of integration step length on accuracy for the shaft with greatest taper. In this problem, four significant figure accuracy can be obtained with about eight steps. A more thorough study of accuracy is made in the next subsection. The approximate buckling loads obtained here for tapered shafts can be compared with the more accurate values presented in Section 4.6.

4.3.4 Accuracy of numerical integration As indicated in Section 4.1.2, the accuracies of numerical integration techniques are affected by both truncation and roundoff errors. The relative importance of these errors is determined by the step length, the nature of the function to be integrated, and the computer word length (see Section 1.2).

As the present buckling problem is not particularly suitable for demonstrating errors, a similar definite integral of a trigonometrical function is considered as follows

$$S = \int_0^{\frac{\pi}{2}} \cos x \, dx = 1 \qquad (4.17)$$

The total error, introduced by either Simpson's rule or the trapezoidal rule, can be expressed as $e = |1 - S|$. The subprogram SIMPV can be used for Simpson's rule, and a very similar one can be written for the trapezoidal rule. Figure 4.8 shows the results obtained using a computer with a word length of sixty bits, giving at least twelve decimal digit precision. The number of integration steps is varied from two to 8192 in multiples of two, and both e and the number of steps are plotted on logarithmic scales.

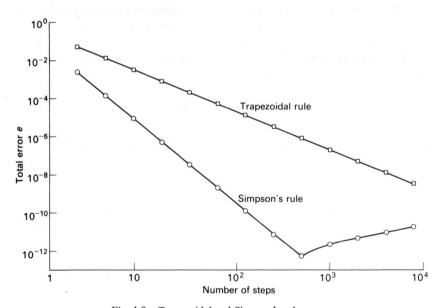

Fig. 4.8   Trapezoidal and Simpson's rule errors

As expected from equations (4.5) and (4.6) for the truncation errors, the Simpson's rule error is very much smaller than the trapezoidal rule error. For Simpson's rule, however, the total error reaches a minimum at 512 steps, after which it is dominated by the roundoff associated with the increasing number of arithmetic operations. It is to be expected that the large trapezoidal rule error will reach a similar minimum, although in the range considered it is still dominated by the truncation error.

It may be concluded that whenever possible Simpson's rule should be used in preference to the trapezoidal rule. Adequate accuracy can often be obtained with as few as ten integration steps, although this depends on the nature of the function involved. There is always a maximum obtainable accuracy and a corresponding optimum number of steps, and these are very dependent on the word length of the computer.

4.4 Numerical Solution of Ordinary Differential Equations

The general analytical solution to the first-order differential equation displayed in equation (4.1) involves an arbitrary constant. A particular solution is then obtained by specifying the value of y at a certain value of x. If an analytical solution is not obtainable, however, a numerical method must be used. As such a method can only give a particular solution, the value of y at one point must be prescribed, and the solution process is normally started from this point.

With higher-order equations, more than one condition must be specified to obtain a particular solution. Consider, for example, the following second-order equation

$$\frac{d^2 y}{dx^2} + a_1(x, y, y') \frac{dy}{dx} + a_0(x, y, y') = 0 \tag{4.18}$$

This may be expressed as two simultaneous first-order equations as follows

$$\frac{dy_1}{dx} = y_2 \tag{4.19}$$

$$\frac{dy_2}{dx} = -a_0(x, y_1, y_2) - a_1(x, y_1, y_2) y_2 \tag{4.20}$$

where $y_1 \equiv y$, $y_2 \equiv y'$. Any ordinary differential equation can be reduced to a set of first-order equations, the number of equations being equal to the order of the highest derivative. To obtain a particular solution to equation (4.18), two conditions must be specified for y or y'. If these conditions are for y and y' at one value of x, the problem is said to be of the initial-value type. A numerical solution process may be started from the point at which y and y', and hence y_1 and y_2 in equations (4.19) and (4.20), are known. If, on the other hand, the conditions for y or y' are specified at different values of x, the problem is said to be of the boundary-value type, for which the method of solution is less straightforward. Similar classifications can be applied to third and high-order equations.

4.4.1 Initial-value problems   There are many practical examples of initial-value problems, such as transient vibrations in mechanical systems and electrical circuits, the solutions to which are determined by the conditions prevailing at the start of the process. Similarly, the trajectory of an unguided rocket is determined by the conditions at launch. Finding the deflection of a cantilevered beam is also an initial-value problem, although it changes to the boundary-value type if both ends of the beam are supported.

The numerical solution of initial-value problems is a subject which has received a considerable amount of attention from numerical analysts, and many different methods have been developed. A detailed discussion of the subject is beyond the scope of this book: the interested reader should consult tests such as Conte (1965), Ralston (1965) or Williams (1972). The approach used here is to introduce one of the more generally useful methods, and to attempt to point out some of the potential difficulties.

4.4.2 The Adams–Bashforth predictor–corrector method   If the function f in equation (4.1) involves both x and y, then it is not possible to prescribe values of y, in order to apply, say, Simpson's rule. One way of overcoming this difficulty is to use a predictor-corrector method. The Adams–Bashforth method is one of the most useful of this type, and is based on the following backward-difference formulae

$$y_{j+1} = y_j + h\left(1 + \frac{1}{2}\nabla + \frac{5}{12}\nabla^2 + \frac{3}{8}\nabla^3 + \ldots\right)f_j \qquad (4.21)$$

$$y_{j+1} = y_j + h\left(1 - \frac{1}{2}\nabla - \frac{1}{12}\nabla^2 - \frac{1}{24}\nabla^3 - \ldots\right)f_{j+1} \qquad (4.22)$$

The difference operator ∇ is such that $\nabla f_j = f_j - f_{j-1}$, and the subscripts and step length, h, have the same meaning as in equation (4.2) and Fig. 4.1. Equation (4.21) is the *predictor* in that it can be used to obtain an estimate for y_{j+1} using only the information available at previous steps. Equation (4.22) is the *corrector* and involves f_{j+1}, which is not known initially. The advantage of the corrector, however, is that it is more accurate, because the coefficients in equation (4.22) are smaller than the corresponding ones in equation (4.21). Therefore, while it is possible to use just a predictor, this is rarely done.

The procedure is to start the solution from the value of x at which y is known, and to progress in suitably small steps. For a typical step from x_j to x_{j+1}, the predictor is used first to estimate y_{j+1} and then the corrector is applied repeatedly, updating f_{j+1} each time, until successive values of y_{j+1} agree to within an acceptable tolerance. If more than one equation is involved, such as equations (4.19) and (4.20), the predictor or corrector is applied to each equation in turn.

In order to be able to use equations (4.21) and (4.22), which involve infinite series of terms, they must be truncated at some point. While accuracy is improved by retaining as many terms as possible, there are some disadvantages which are discussed below. Truncating after the first difference of derivatives, the predictor and corrector become

$$y_{j+1} = y_j + h\left(1 + \frac{1}{2}\nabla\right)f_j + e_\mathrm{p} \tag{4.23}$$

$$y_{j+1} = y_j + h\left(1 - \frac{1}{2}\nabla\right)f_{j+1} + e_\mathrm{c} \tag{4.24}$$

where the truncation errors are given by

$$e_\mathrm{p} \approx 5h\,\nabla^2 f_j/12, \quad e_\mathrm{c} \approx -\,h\,\nabla^2 f_{j+1}/12 \tag{4.25}$$

An advantage of using both a predictor and a corrector is that, since $e_\mathrm{p} \approx -5e_\mathrm{c}$, an estimate of accuracy can be obtained. If Y is the true value of y_{j+1}, Y_p the predicted value, and Y_c the corrected value after convergence, then

$$Y = Y_\mathrm{p} - 5e_\mathrm{c} = Y_\mathrm{c} + e_\mathrm{c} \tag{4.26}$$

and the truncation error of the computed solution is

$$e_\mathrm{c} \approx \frac{1}{6}(Y_\mathrm{p} - Y_\mathrm{c}) \tag{4.27}$$

A disadvantage of predictor–corrector methods is that they are usually not self-starting. In order to use equation (4.23), the value of f_{j-1} is required, but is not available for the first step of the process. The same restriction does not apply, however, to the corrector, equation (4.24). For present purposes it is sufficient to truncate the predictor by one more term for the first step only. More accurate schemes, which retain more terms and therefore need information from more previous points, require special methods for the first few steps. Self-starting methods such as the Runge–Kutta type are appropriate for this purpose. Indeed, Runge–Kutta methods are widely used for the complete solution process. Their disadvantages are that they generally involve rather more computation for the same accuracy as predictor–corrector methods, and that it is much more difficult to estimate this accuracy.

It is important to establish that the repeated application of the corrector is a convergent process. Convergence is normally assured if the step length is small. The most useful criterion for choosing h is such that about two cycles of iteration are required for convergence. If h is increased, the reduction in the number of steps is offset by the increase in the number of iterations required at each step, and convergence may not be achieved. If h is reduced, however, somewhat fewer iterations but more steps are needed. As convergence after only two cycles implies a high rate of convergence, the choice of tolerance is not very critical. The optimum step length may vary significantly as the solution proceeds, and ideally provision should be made for changing h.

The accuracy of the final solution depends on the truncation error of the corrector, and to a lesser extent on roundoff errors. Since the method of solution is of a step-by-step type, it is necessary to consider not only the magnitudes of these errors introduced at each step, but also the extent to which they grow as the solution proceeds. If this growth is rapid, the process is said to be unstable. Such instability may be inherent in the particular differential equation, but can also be introduced by the method of solution. The requirements for stability and accuracy are often conflicting. Suffice it to say that the Adams–Bashforth method has good stability properties.

4.4.3 Boundary-value problems Typical boundary-value problems involve second-order differential equations, with conditions for the dependent variable or its first derivative specified at two values of the independent variable. Problems of this type can be solved by finite difference methods, as described in Section 6.3.

In principle, it is possible to solve all boundary-value problems by the initial-value process described above. The main difficulty is that one boundary condition is known at each end of the range of solution, whereas two conditions are required at one end to start the initial-value process. The usual way of overcoming this is to employ a 'shooting' method. The second initial condition at one end is guessed, and the corresponding solution obtained as far as the second boundary. An iterative procedure is set up to adjust the guessed value until the condition at this boundary is satisfied to within an acceptable tolerance. Shooting methods are difficult to apply if the solution process tends to be inherently unstable. The use of a shooting method is demonstrated in Section 4.6, although the problem considered is not quite of the boundary-value type discussed here.

4.5   Case Study: Railway Train Performance

This example serves to illustrate not only the solution of a typical initial-value problem, but also the use of data interpolation and curve fitting, which were introduced in Chapter 3. It also demonstrates the use of BLOCK DATA subprograms and EXTERNAL statements. Although SI units are employed elsewhere in this book, the data used here are in British units and these are retained in the calculations.

Figure 4.9 shows the forces acting on a railway locomotive and carriages travelling up a slope of angle θ. The equation of motion parallel to the slope is

$$\frac{(W_L + W_C)}{g_0} \frac{d^2y}{dt^2} = E - D - (W_L + W_C) \sin \theta \qquad (4.28)$$

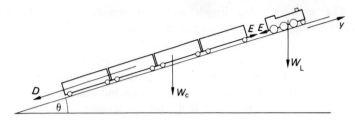

Fig. 4.9   Forces acting on a railway train

where y is the distance travelled in time t, and W_L and W_C are the weight of the locomotive and carriages respectively. The force E is the drawbar tractive effort (DBTE) applied by the locomotive to the carriages, and is obtained from dynamometer measurements. It therefore allows for drag forces on the locomotive, but not for gravitational forces due to the slope. The force D is the drag on the carriages alone. Both E and D are functions of speed, and θ varies with y. If the position and velocity of the train are specified at a particular time, the subsequent motion can be computed numerically.

4.5.1 Problem specification   The performance of Merchant Navy Class express passenger steam locomotive Number 35020 between London and Salisbury is to be predicted, and compared with timed runs. The main reason for selecting this somewhat out-of-date example is that a very comprehensive set of data has been published by the British Transport Commission (1958). The method is equally applicable to more modern locomotives and rolling stock.

The locomotive steaming rate is to be taken as 24000 lb/hr, about the maximum that can be maintained by one man stoking continuously. At the start of the run the steaming rate is 20000 lb/hr, increasing linearly with time to the full rate over the first ten minutes. Because of speed restrictions, the steaming rate is also reduced to 20000 lb/hr between mileposts 55 and 66. The effect of such reductions can be approximated by assuming that the DBTE is decreased in proportion to the steaming rate from its value at 24000 lb/hr for the same speed. There is a general speed limit of 85 mile/hr, and one of 40 mile/hr over the first 3·9 mile. The weight of the locomotive is 151·2 ton, and the calculation is to be performed for a total weight of carriages and passengers of 400 ton, equivalent to eleven fully laden carriages.

4.5.2 Solution   Before considering a program for this problem, it is necessary to decide how the various types of data are to be handled. The slope of the track can be obtained from the gradient profile for the route. If the gradient is '1 in G' between two particular points, then

$$\sin \theta = 1/G \tag{4.29}$$

for this length of track. The required gradient data may be stored for computing purposes as values of G, together with the positions at which the corresponding stretches of track start. In the case of level track, G is taken to be 10^8, a suitably large number.

The drag force on the carriages is available as a curve of specific resistance against speed. Since this resistance is due to a combination of mechanical friction and wind resistance, it is likely to be a quadratic function of speed. An extremely good fit can be obtained with the following relationship

$$R = 2.1 + 0.05V + 0.001V^2 \text{ lb/ton} \tag{4.30}$$

where V is speed in mile/hr, and $D = RW_C$. On the other hand, the curve of DBTE against speed for a steaming rate of 24000 lb/hr is much less amenable to simple curve fitting. It is therefore preferable to store the coordinates of a series of points on the curve and to interpolate, as discussed in Section 3.6.4.

The most suitable set of units to use in the calculation is that of tons, miles and hours. Hence, in equation (4.28), $g_0 = 78973$. Since the unit of DBTE is the ton, the only conversion required is that of D from lb to ton.

Figures 4.10 to 4.13 show a main program and three subprograms for solving this problem. In view of their complexity, it is convenient to consider them under separate headings.

4.5.3 The main program The main program shown in Fig. 4.10 serves to read in data not already stored elsewhere, to set the initial conditions and control the stepping of the predictor-corrector process, and to write out the results. The subscripted variables Y and F are employed to store the functions and derivatives used in the predictor-corrector process as described below. The variables DIST, SPEED, ACCLN and GRAD are easily recognised as distance, speed, acceleration and gradient. Similarly, WCARR, WTOTAL and DBTE are used for W_C, $(W_L + W_C)$ and E. The integer IGRAD is the current subscript of the gradient data array, the use of which is described in Section 4.5.5. All of these non-subscripted variables are located in the COMMON block of storage named CVAR.

The main program first reads a title for the particular problem from two data cards into the subscripted variable TITLE. Any combination of alphanumeric characters can be stored by this means, as indicated in Section 1.2.3. From the next card are read the train weight data (the locomotive weight is stored in WLOCO), route distance RDIST (for terminating the computation), time step length HTIME for the numerical solution and the frequency IFREQ with which results are to be written out. If IFREQ is read in as ten, for example, results will be written out after every ten time steps. These data are immediately written out with a program title and suitable captions.

Before the solution procedure is started, the necessary preliminary calculations are performed, and a heading prepared for the subsequent table of

```
C   PROGRAM TO COMPUTE RAILWAY TRAIN PERFORMANCE.
C
        DIMENSION Y(2000,2),F(2000,2),TITLE(16)
        COMMON  /CVAR/ DIST,SPEED,ACCLN,GRAD,IGRAD,WCARR,WTOTAL,DBTE
        EXTERNAL  FNS
C
C   READ MAIN TITLE, WEIGHT OF LOCOMOTIVE, WEIGHT OF CARRIAGES AND
C   CONTENTS, TOTAL DISTANCE OF ROUTE, TIME INTERVAL FOR THE NUMERICAL
C   SOLUTION, AND FREQUENCY OF PRINTING RESULTS.
        READ(5,51) TITLE,WLOCO,WCARR,RDIST,HTIME,IFREQ
   51   FORMAT(8A10/8A10/4F10.0,I5)
        WRITE(6,61) TITLE,WLOCO,WCARR,RDIST,HTIME
   61   FORMAT(34H1RAILWAY TRAIN PERFORMANCE PROGRAM / 1X,8A10/ 1X,8A10//
       1 23H WEIGHT OF LOCOMOTIVE =,F7.1,5H TONS /
       2 35H WEIGHT OF CARRIAGES AND CONTENTS =,F7.1,5H TONS /
       3 30H TOTAL DISTANCE OF THE ROUTE =,F7.1,6H MILES /
       4 39H TIME INTERVAL FOR NUMERICAL SOLUTION =,E12.3,5H HOUR )
C
C   CALCULATE TOTAL WEIGHT, SET INITIAL CONDITIONS AND WRITE HEADING.
        WTOTAL=WLOCO+WCARR
        Y(1,1)=0.
        Y(1,2)=0.
        CALL  FNS(Y,F,0.,1)
        IGRAD=1
        NITER=0
        WRITE(6,62)
   62   FORMAT(61H0    MINS      MILES       MPH      ACCLN       DBTE
       1 GRAD)
C
C   SET UP THE PREDICTOR-CORRECTOR SOLUTION LOOP.
        DO 1 ITIME=2,2000
        TIME=HTIME*FLOAT(ITIME-1)
        CALL   PRECOR(Y,F,TIME,2,2000,ITIME,HTIME,ITER,FNS)
C
C   ACCUMULATE TOTAL NUMBER OF CORRECTOR ITERATIONS.
        NITER=NITER+ITER
        IF(MOD(ITIME,IFREQ).NE.0.AND.DIST.LT.RDIST) GO TO 1
C
C   OUTPUT RESULTS AT REQUIRED TIME INTERVALS.
        TMINS=TIME*60.
        IF(ABS(GRAD).LT.1.E4) WRITE(6,63) TMINS,DIST,SPEED,ACCLN,DBTE,GRAD
        IF(ABS(GRAD).GE.1.E4) WRITE(6,64) TMINS,DIST,SPEED,ACCLN,DBTE
   63   FORMAT(1X,5F10.2,F10.0)
   64   FORMAT(1X,5F10.2,5X,5HLEVEL)
C
C   LEAVE SOLUTION LOOP WHEN ROUTE COMPLETED.
        IF(DIST.GE.RDIST) GO TO 2
   1    CONTINUE
C
C   NORMAL EXIT FROM SOLUTION LOOP INDICATES MAXIMUM SUBSCRIPT REACHED.
        WRITE(6,65)
   65   FORMAT(41H0EXECUTION TERMINATED BY INDEX SIZE LIMIT)
        STOP
C
C   FIND AVERAGE CORRECTOR ITERATIONS AND ACCURATE JOURNEY TIME.
   2    CYCLES=FLOAT(NITER)/FLOAT(ITIME-1)
        TIME=TIME-HTIME*(Y(ITIME,1)-RDIST)/(Y(ITIME,1)-Y(ITIME-1,1))
        TMINS=TIME*60.
C
C   OUTPUT THE RESULTS.
        WRITE(6,66) TMINS,NITER,CYCLES
   66   FORMAT(21H0TOTAL JOURNEY TIME =,F7.2,8H MINUTES /
       1 29H TOTAL CORRECTOR ITERATIONS =,I5,5X,16H MEAN PER STEP =,F5.2)
        STOP
        END
```

Fig. 4.10   Main program for railway train performance calculation

results. These preliminaries include prescribing the initial values for distance, velocity and their derivatives, and setting NITER, the counter for the number of corrector iterations, to zero. The predictor-corrector process is stepped, with time (stored in the program variable TIME) as the independent variable. Note that if 2000 steps are completed without the end of the route being reached, execution is terminated to avoid exceeding the dimensions of the Y and F arrays. Having set the current time for each step, the predictor-corrector subprogram PRECOR is called. When control of execution returns to the main program, the number of corrector iterations for the step, stored in the argument ITER, is added to the current total. If the step counter ITIME is not a multiple of IFREQ, and the end of the route has not been reached, the computation proceeds to the next step. Otherwise, the current values of time (in minutes) and the other relevant quantities are written out. Two alternative output statements are provided so that large values of GRAD, corresponding to level track, are replaced by the label 'LEVEL' in the printed results.

As soon as the distance travelled exceeds the route distance, execution is transferred out of the predictor-corrector stepping loop. The average number of corrector iterations per step is found and stored in the variable CYCLES, and a linear interpolation performed to find more accurately the time at which the train passed the end of the route during the last time step. These results are written out and execution is then terminated.

4.5.4 The predictor-corrector subprogram

Figure 4.11 shows a SUBROUTINE subprogram named PRECOR for the Adams–Bashforth method of solving initial-value type problems, as described in Section 4.4.2. The subprogram is written in such a way as to be applicable to a wide variety of problems. Thus, the variables Y and F, and in particular their maximum array sizes, are all entered as arguments, as described in Section 4.2.3. NEQN defines the number of simultaneous first-order ordinary differential equations to be solved, while NPT specifies the maximum number of points permitted in the solution stepping procedure.

The variable names are very similar to those introduced in equation (4.1): X, Y and F are equivalent to x, y and f respectively. As equation (4.28) for the present problem is second-order, it must be reduced to two first-order equations by the method described in Section 4.4. Comparing equation (4.28) with equation (4.18), the required equations equivalent to equations (4.19) and (4.20) are

$$\frac{dy_1}{dt} = f_1 = y_2 \tag{4.31}$$

$$\frac{dy_2}{dt} = f_2 = g_0 \frac{(E - D - (W_C + W_L)/G)}{(W_C + W_L)} \tag{4.32}$$

```
      SUBROUTINE   PRECOR(Y,F,X,NEQN,NPT,JPT,H,KMAX,FNCALL)
C
C  SUBROUTINE FOR NUMERICAL SOLUTION OF ORDINARY DIFFERENTIAL EQUATIONS
C  BY ADAMS-BASHFORTH PREDICTOR-CORRECTOR METHOD, TRUNCATING AFTER FIRST
C  DIFFERENCES OF DERIVATIVES.
C
      DIMENSION  Y(NPT,NEQN),F(NPT,NEQN)
      IPT=JPT
      IF(IPT.GE.2.AND.IPT.LE.NPT) GO TO 1
      WRITE(6,61) IPT
   61 FORMAT(6HOIPT =,I5,5X,23HINVALID INDEX IN PRECOR)
      STOP
C
C  APPLY PREDICTOR TO EACH EQUATION IN TURN.
    1    DO 2 IEQN=1,NEQN
      DELF=0.
      IF(IPT.GE.3) DELF=F(IPT-1,IEQN)-F(IPT-2,IEQN)
    2 Y(IPT,IEQN)=Y(IPT-1,IEQN)+H*(F(IPT-1,IEQN)+0.5*DELF)
C
C  DEFINE CONVERGENCE TOLERANCE AND MAXIMUM NUMBER OF ITERATIONS.
      TOLER=1.E-6
      NCYCLE=20
C
C  SET UP THE CORRECTOR ITERATION LOOP.
      DO 4 K=1,NCYCLE
C
C  EVALUATE THE DERIVATIVES.
      CALL FNCALL(Y,F,X,IPT)
C
C  APPLY THE CORRECTOR.
      SUMERR=0.
      DO 3 IEQN=1,NEQN
      DELF=F(IPT,IEQN)-F(IPT-1,IEQN)
      YTEMP=Y(IPT-1,IEQN)+H*(F(IPT,IEQN)-0.5*DELF)
      ERR=YTEMP-Y(IPT,IEQN)
      IF(YTEMP.NE.0.) ERR=ERR/YTEMP
      Y(IPT,IEQN)=YTEMP
    3 SUMERR=SUMERR+ABS(ERR)
C
C  TEST FOR CONVERGENCE.
      IF(SUMERR.LT.TOLER) GO TO 5
    4 CONTINUE
C
C  NORMAL EXIT FROM ITERATION LOOP INDICATES FAILURE TO CONVERGE.
      K=NCYCLE
      WRITE(6,62) NCYCLE
   62 FORMAT(36H CONVERGENCE FAILURE IN PRECOR AFTER,I3,7H CYCLES)
    5 KMAX=K
      RETURN
      END
```

Fig. 4.11   Subprogram for Adams-Bashforth predictor-corrector method

The independent variable, represented by X in PRECOR, is time in the physical problem. Distance and velocity are represented by y_1 and y_2 respectively, where the subscripts here refer to equation numbers rather than point numbers as in Section 4.4.2. Hence, Y and F in PRECOR have two subscripts, typically Y(IPT, IEQN). IPT is equivalent to $j + 1$ in equations (4.23) and (4.24), while IEQN is either one or two according to which of the simultaneous equations, (4.31) or (4.32), is being used. The maximum value of IPT is arbitrarily set at 2000 by the DIMENSION statement in the main program.

The first five arguments of PRECOR have already been discussed. JPT enters the point associated with the current step, and its value is assigned to IPT in the first executable statement. This is a programming refinement to reduce execution time, as outlined in Section 2.6.1. Every time a subprogram argument is used in an arithmetic expression its value must be obtained from the calling program, which takes significantly longer than when the variable is local to the subprogram. Therefore, the value of the local variable IPT, which is used repeatedly, is defined at the beginning of the subprogram, and JPT is used only once. This refinement is not normally applicable to subscripted arguments. The argument H defines the step length for X, and is equivalent to h in equations (4.23) and (4.24). The variable KMAX is used to return the maximum value of the counter for the corrector iterations, K, to the calling program. The final argument, FNCALL, is the dummy name for the subprogram called by PRECOR to define the functional forms of the differential equations being solved.

While the names of the arguments in PRECOR are appropriate for the general problem amenable to the predictor–corrector method, the equivalent values and variables for the present problem are displayed by the relevant CALL statement in the main program, Fig. 4.10. Note in particular that the subprogram name, FNS, is now the actual name of the subprogram shown in Fig. 4.12. Because FNS is a subprogram name used as an argument for another subprogram, it must be declared in an EXTERNAL statement at the beginning of the calling program.

Returning attention to Fig. 4.11, the validity of the step counter is first tested. Then the predictor, equation (4.23), is applied to each simultaneous differential equation in turn. The backward difference DELF, which is equivalent to ∇f_j in equation (4.23), is only defined if there are sufficient previous values of f, otherwise it is set to zero. As discussed in Section 4.4.2, the predictor is not self-starting, and is truncated for the first step of the solution process. An iteration loop is then set up for the corrector. The maximum number of cycles, NCYCLE, is defined as twenty, which is ample for a process which is expected to converge in about two cycles. The convergence tolerance, TOLER, is defined as 10^{-6}, which reflects the level of accuracy to be expected from the present method. The subprogram could be further generalised by treating both NCYCLE and TOLER as arguments, and supplying appropriate values from the calling program.

The first action within the corrector iteration loop is to define the derivatives, F, at the current point in the range of the independent variable, using the current values of the dependent variables Y. The derivatives are defined in a subprogram, whose name FNCALL is a dummy for the actual name, FNS, supplied by the main program. Then the corrector, equation (4.24), is applied to each differential equation in turn, and the convergence error, ERR, is defined as the relative change in the particular dependent variable between successive cycles of iteration. The sum of the magnitudes of the errors for all the differential equations is accumulated in SUMERR. When this sum is less than the prescribed tolerance, the correction process is terminated, the number of iterations is assigned to the argument KMAX, and control of execution returns to the calling program. As with the argument JPT, the use of KMAX within the subprogram is restricted to one statement. If the correction process has not converged within the prescribed maximum number of cycles, a warning message is written out, but execution is not terminated.

The final corrected values of the dependent variables and derivatives at each step remain stored in the arrays Y and F, and are therefore available for use in later steps. It should be noted, however, that both the predictor and corrector use the values at the initial point IPT = 1. Hence, in the main program it is necessary to define not only the initial conditions for the dependent variables, but also the corresponding derivatives, by calling FNS.

While the predictor–corrector subprogram shown in Fig. 4.11 is reasonably general, many further refinements are possible. For example, estimates of the truncation error at each step could be computed with the aid of equation (4.27). Although more accurate results could be obtained by retaining more terms in equations (4.21) and (4.22), special methods would then be needed to start the process. Despite its relative simplicity, PRECOR is a very useful subprogram for the Adams–Bashforth method, and can be used to solve a wide variety of problems.

4.5.5 The subprograms for the derivatives   Figure 4.12 shows a subprogram named FNS for specifying the functional forms of the differential equations defined in equations (4.31) and (4.32). FNS is normally called from PRECOR by the dummy name FNCALL, although it is also called from the main program to evaluate the derivatives at the initial point of the solution. As this subprogram is coded for the specific problem being considered, there is no need to allow the dimensions of the Y and F arrays to be variable. They must, however, agree with the dimensions defined in the main program.

Data from the main program required for evaluating the derivatives are stored in the COMMON block of storage named CVAR. As indicated in Section 4.5.2, further data are required to define the variation of locomotive DBTE with speed for the appropriate steaming rate, and the variation of track gradient with distance along the route. The DBTE data are stored in the array TEDATA, as a series of nineteen values for speeds of 0, 5, 10, . . . , 90 mile/hr. As these speeds are uniformly spaced, there is no need to store them explicitly. For the track-gradient data, however, it is necessary to store both the gradients and the distances along the route at which the changes to those gradients are made. These quantities are stored in the arrays GDATA and GCDIST respectively. The tractive effort and gradient data could be read into the relevant arrays from data cards, during execution of the program. Since they are unchanged for a particular route and steaming rate, however, it is more convenient to use the DATA statement facility to enter them during compilation of the program. On the other hand, if the program is to be applicable to other routes and steaming rates, the data would have to be changed. A suitable compromise is to locate the arrays in COMMON blocks of storage and to enter the data with the aid of a separate BLOCK DATA subprogram.

```
      SUBROUTINE  FNS(Y,F,TIME,JPT)
C
C  SUBROUTINE FOR DEFINING THE FUNCTIONAL FORMS OF THE DERIVATIVES.
C
      DIMENSION  Y(2000,2),F(2000,2)
      COMMON  /CTDATA/ TEDATA(19)  /GCDATA/ NGC,GCDIST(100),GDATA(100)
     1          /CVAR/ DIST,SPEED,ACCLN,GRAD,IGRAD,WCARR,WTOTAL,DBTE
      IPT=JPT
      DIST=Y(IPT,1)
      SPEED=Y(IPT,2)
C
C  IMPOSE SPEED RESTRICTIONS.
      IREST=0
      IF(DIST.GT.3.9) GO TO 1
      IF(SPEED.LE.40.) GO TO 2
      SPEED=40.
      IREST=1
      GO TO 2
    1 IF(SPEED.LE.85.) GO TO 2
      SPEED=85.
      IREST=1
C
C  FIND CURRENT GRADIENT ACCORDING TO DISTANCE ALONG THE ROUTE.
    2 NGCM1=NGC-1
      DO 3 ISCAN=IGRAD,NGCM1
      IF(DIST.GE.GCDIST(ISCAN).AND.DIST.LT.GCDIST(ISCAN+1)) GO TO 4
    3 CONTINUE
      ISCAN=NGC
    4 IGRAD=ISCAN
      GRAD=GDATA(IGRAD)
C
C  CALCULATE DRAWBAR TRACTIVE EFFORT FROM STORED PERFORMANCE CURVE.
      IV=SPEED/5.+1.
      DEL=AMOD(SPEED,5.)/5.
      DBTE=TEDATA(IV)+DEL*(TEDATA(IV+1)-TEDATA(IV))
C
C  CALCULATE SPECIAL TRACTIVE EFFORT REDUCTIONS.
      FACTOR=1.
      IF(DIST.GT.55..AND.DIST.LT.66.) FACTOR=0.8
      IF(TIME.LT.0.167) FACTOR=0.8+0.2*TIME/0.167
      DBTE=DBTE*FACTOR
C
C  CALCULATE SPECIFIC RESISTANCE ACCORDING TO SPEED, THEN THE ACCELERATION.
      RESIST=SPEED*(0.001*SPEED+0.05)+2.1
      ACCLN=78973.*(DBTE-RESIST*WCARR/2240.-WTOTAL/GRAD)/WTOTAL
      IF(IREST.NE.0.AND.ACCLN.GT.0.) ACCLN=0.
C
C  FINALLY DEFINE THE DERIVATIVES.
      F(IPT,1)=SPEED
      F(IPT,2)=ACCLN
      RETURN
      END
```

Fig. 4.12   Subprogram for defining derivatives

Figure 4.13 shows such a subprogram, which contains only COMMON and DATA statements. The necessary dimensioning of the arrays can be accomplished in the COMMON statement. The tractive effort data array is located in the block named CTDATA, while the gradient data arrays are in GCDATA. Although the dimensions of the latter arrays allow for up to a hundred changes of gradient, only eighty are involved in the present problem. This actual number of changes is stored in the variable NGC, for use in subprogram FNS. In a BLOCK DATA subprogram, it is not possible to enter only part of an array. Therefore, the array names appear unsubscripted in the

```
BLOCK   DATA
COMMON  /CTDATA/ TEDATA(19)   /GCDATA/ NGC,GCDIST(100),GDATA(100)
DATA   TEDATA,NGC,GCDIST,GDATA   /
1       19.0,16.6,13.9,11.5,9.60,8.12,6.95,5.97,5.26,4.64,
2       4.11,3.57,3.17,2.72,2.32,1.96,1.56,1.25,0.90,   80,
3       0.  ,0.20,0.25,0.55,1.35,1.70,1.90,2.40,2.50,3.20,5.00,5.45,
4       6.20,6.70,7.00,7.40,11.1,11.6,12.2,12.6,13.2,13.7,13.8,13.9,
5       14.1,15.2,16.0,16.4,17.0,18.6,18.8,19.9,20.1,20.5,23.7,26.2,
6       29.3,30.2,31.0,31.2,31.7,31.8,35.5,36.2,38.2,40.3,41.5,41.9,
7       42.8,43.8,44.0,44.3,45.1,45.3,45.9,46.3,47.7,48.2,51.1,52.7,
8       53.5,57.2,57.7,58.6,59.3,60.0,61.3,61.5,62.7,65.7,66.3,67.9,
9       68.9,70.7,73.5,74.3,76.1,77.8,79.2,81.6,20*0.,
1        726., 141.,-226.,  1.E8,-199.,  250.,-234.,  445., 1.E8, 338.,
2       -798.,  1.E8, 347.,  448.,-741.,  1.E8, 373.,-378.,  1.E8,-361.,
3        1.E8, 348.,  1.E8,-506.,  1.E8, 529.,  1.E8, 446.,1334.,-460.,
4       -330.,-547.,  1.E8, 387.,  326.,  314.,  298.,  304.,-444.,-315.,
5       -497.,  1.E8, 655.,  1.E8, 337.,  1.E8, 785.,  480.,-368.,-768.,
6        1.E8, 385.,  694.,  1.E8, 393.,  249.,  434.,  249.,  1.E8, 287.,
7       -550.,  1.E8,-345.,-560.,  1.E8,-194.,  1.E8, 287.,-178.,  1.E8,
8        220.,-330.,  264.,  165.,-440.,-735.,-245.,-140.,-169.,  1.E8,
9       20*0.  /
END
```

Fig. 4.13   Subprogram for entering numerical data during program compilation

DATA statement, followed by the 220 corresponding items of data. The nineteen values of DBTE range from 19.0 ton at rest to 0.90 ton at 90 mile/hr. Then NGC is set to eighty, and the eighty distances of the gradient changes along the route are stored in GCDATA, the last twenty elements of which are set to zero. Finally, the eighty values of gradient are stored, followed by twenty zeros to fill the array GDATA. Negative gradients mean downhill slopes, while values of 10^8 imply level track. This stored information is to be interpreted as follows. At the beginning of the route, the gradient is 1 in 726 up, changing to 1 in 141 up after 0·20 miles, and so on. At the other end of the range, the track is level after 81·6 mile, which is actually slightly beyond the end of the route. The advantage of using such a BLOCK DATA subprogram is that a new version can be substituted if, for example, a new route is to be studied, but without disturbing the rest of the program.

Returning attention to Fig. 4.12, the value of the argument JPT is first assigned to the local variable IPT, as in PRECOR. For the same reason, the current values of distance and speed are assigned to the variables DIST and SPEED. Apart from reducing execution time, this also serves to make the coding easier to read. Since these two variables are located in the COMMON block of storage, CVAR, the values stored in FNS are also available in the main program. The rest of the subprogram is concerned with establishing the relationships for the derivatives defined in equations (4.31) and (4.32), and applying the additional constraints described in Section 4.5.1, the problem specification.

Speed restrictions are considered first, both the general one of 85 mile/hr and the special one of 40 mile/hr over the first 3·9 mile. In addition to restricting the speed, however, it is also necessary to avoid a positive acceleration when running at a maximum speed. This is done with the aid of the variable IREST, which is normally zero, but is set to one when a speed restriction is imposed, and is used later in the subprogram.

The current track gradient is found by scanning the stored data until the value of DIST lies between successive distances for gradient changes. The index limits for this scanning process are set as IGRAD, the value at the last time step, and NGCM1, which is one less than the number of stored distances. If DIST exceeds the last stored distance, the last gradient is used. The initial value of IGRAD is set as one in the main program.

The DBTE is found by linear interpolation in the TEDATA array. This is done in much the same way as in the case study described in Section 3.6, although the constant 5 mile/hr speed intervals allow the coding to be made particularly concise. The required reductions of DBTE over the first ten minutes and between mileposts 55 and 66 are then applied.

The specific resistance of the carriages is then calculated with the aid of equation (4.30). Note the rearrangement of this quadratic expression to minimise the number of arithmetic operations. The acceleration is defined using equation (4.32), but is set to zero if there is a tendency to exceed one of the speed restrictions. Finally, the values of the two derivatives defined in equations (4.31) and (4.32) are assigned to the relevant elements of the array F.

4.5.6 The overall program As this relatively complex program has been discussed in terms of its individual subprograms, some general comments are now appropriate. These concern the organisation of the overall program, and some of the differences between the various types of subprogram.

Figure 4.14 shows the effective layout of the program, with subprograms on the left and COMMON blocks of storage on the right. All of these occupy quite distinct regions of the computer core store, and communications between them

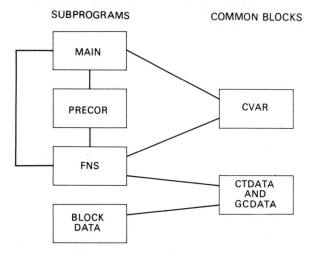

Fig. 4.14   General layout of overall railway train performance program

are indicated by the connecting lines. Thus, variables located in the block CVAR are accessible by the main program, and by FNS, while the variables in the other blocks are only available for computations in FNS. The subprogram PRECOR communicates with the main program and with FNS by means of arguments only. The direct communication between the main program and FNS is for the purpose of defining the initial values of the derivatives. The BLOCK DATA subprogram is only used during compilation of the program, to store data in the COMMON blocks.

Variables not located in COMMON blocks are effectively stored in the same region of the core as the subprogram in which they first appear. Thus, the arrays Y and F, for example, are introduced in the main program, and are carried through as arguments into both PRECOR and FNS. When they are used in the latter subprogram, the relevant values must be sought by tracing the calling sequence back to the main program. Consequently, the use of arguments for communications between subprograms is relatively wasteful of execution time compared to using COMMON blocks. The advantage of arguments is that they permit very general subprograms to be written. For example, PRECOR, which relies entirely on arguments, is independent of the particular problem to which it is applied. For the same reason, the variable names used in PRECOR are the general ones employed in Section 4.4.2, whereas the corresponding names used in the other subprograms are more readily associated with the particular problem.

4.5.7 Results and discussion Figure 4.15 shows the results obtained with a time step length of 0·002 hr and IFREQ = 20. More detailed output could be obtained by reducing IFREQ. At 2·4 min intervals, the distance travelled in miles, speed in mile/hr, acceleration in mile/hr^2, DBTE in ton, and track gradient are all printed out. At three particular steps in the solution, the corrector fails to converge after twenty cycles of iteration. Closer inspection shows that these failures are associated with imposed speed restrictions, which create discontinuities in the computed derivatives. The results are not affected, however, because the speed and acceleration are correctly defined. It should be noted that the printed DBTE values have not been reduced in regions of speed restrictions, as must be done in practice.

At the end of the results, the journey time is printed, together with the total number of corrector iterations and the mean number per step of the solution. A step length of 0·002 hr would appear to be a reasonable choice, on the basis of the mean number of corrector iterations (which is not significantly affected by the three convergence failures). If the program is re-run with various step lengths, the results shown in Table 4.1 are obtained. Owing to the discontinuous nature of the track gradient, this is not a good example to illustrate the optimum choice of step length for an initial-value problem. Table 4.1 shows that the journey time is not significantly affected by the choice of step length. A value

RAILWAY TRAIN PERFORMANCE PROGRAM
ROUTE FROM LONDON (WATERLOO) TO SALISBURY
LOCOMOTIVE NUMBER 35020 STEAMING AT 24000 LB/HR

WEIGHT OF LOCOMOTIVE = 151.2 TONS
WEIGHT OF CARRIAGES AND CONTENTS = 400.0 TONS
TOTAL DISTANCE OF THE ROUTE = 81.5 MILES
TIME INTERVAL FOR NUMERICAL SOLUTION = 0.200E-02 HOUR

MINS	MILES	MPH	ACCLN	DBTE	GRAD
2.28	0.88	39.00	511.77	4.57	LEVEL
CONVERGENCE FAILURE IN PRECOR AFTER 20 CYCLES					
4.68	2.48	40.00	0.00	4.70	445.
7.08	4.08	42.03	289.12	4.71	338.
9.48	6.00	54.89	306.54	3.54	LEVEL
11.88	8.34	62.89	181.56	2.91	LEVEL
14.28	10.97	68.17	93.55	2.47	LEVEL
16.68	13.78	72.65	-203.56	2.13	348.
19.08	16.68	71.63	-138.17	2.20	446.
21.48	19.54	74.10	240.45	2.02	-330.
23.88	22.48	70.46	-147.33	2.29	387.
26.28	25.20	65.38	-103.65	2.69	326.
28.68	27.74	62.19	-57.88	2.97	314.
31.08	30.19	60.35	-39.39	3.14	298.
33.48	32.71	67.13	110.38	2.55	LEVEL
35.88	35.47	70.38	57.99	2.29	LEVEL
38.28	38.29	71.17	-188.33	2.24	337.
40.68	41.03	67.95	97.06	2.48	LEVEL
43.08	43.78	71.89	249.52	2.18	-368.
45.48	46.62	68.78	-233.57	2.42	249.
47.88	49.24	62.52	-129.10	2.94	249.
50.28	51.67	61.43	206.94	3.04	LEVEL
52.68	54.22	66.37	266.24	2.61	-550.
55.08	57.02	72.75	104.57	1.70	-550.
57.48	59.99	74.94	-68.13	1.57	LEVEL
59.88	63.06	75.52	367.27	1.53	-178.
CONVERGENCE FAILURE IN PRECOR AFTER 20 CYCLES					
62.28	66.32	83.63	-507.41	1.33	220.
64.68	69.42	74.78	-308.54	1.98	264.
67.08	72.16	62.16	-284.48	2.98	165.
69.48	74.55	63.39	280.29	2.86	-735.
71.88	77.30	74.81	312.51	1.97	-245.
CONVERGENCE FAILURE IN PRECOR AFTER 20 CYCLES					
74.28	80.56	85.00	0.00	1.25	-169.
75.00	81.58	85.00	0.00	1.25	-169.

TOTAL JOURNEY TIME = 74.94 MINUTES
TOTAL CORRECTOR ITERATIONS = 1421   MEAN PER STEP = 2.27

Fig. 4.15   Results from railway train performance program

which gives an appropriate degree of detail along the route should therefore be selected.

With such a complex program, very thorough testing is essential, as discussed in Section 2.5. The first check should be on the values of gradient and DBTE printed out for particular distances and speeds. One or two accelerations can then be calculated manually, from which the changes in speed and distance over small time intervals can be found and compared. In effect, this procedure checks that the computed solution satisfies the original differential equation. A further test can be devised by simplifying the problem until it is amenable to analytical solution. For example, both the gradient and DBTE can be made constant, and the quadratic term eliminated from equation (4.30). When this is done, the

Table 4.1
Results for various time step lengths

Step length (hr)	Total iterations	Mean iterations per step	Journey time (min)
0·001	1725	1·38	74·90
0·002	1421	2·27	74·94
0·005	881	3·54	74·47
0·01	660	5·28	74·55
0·02	440	6·98	74·55

computed results compare very favourably with the appropriate analytical solution.

The results shown in Fig. 4.15 can be compared with actual performance on timed runs. The total route distance of 81·5 mile is the distance at which braking would normally be applied in order to stop at Salisbury after 83·7 mile. The time allowed by the timetable was 80 min to Salisbury, implying about 76 min to milepost $81\frac{1}{2}$, which is in good agreement with the computed result. More detailed comparisons with timed runs show similar agreement for passing times and speeds at various points along the route. The only discrepancies appear to be due to crew psychological factors. The speeds in practice are higher on the uphill stretches of track and lower than predicted on the downhill ones. These differences are due to increased or reduced steaming rates.

A program of the present type for railway-train performance has many practical applications. For example, it can be used to aid timetabling, in particular the effect of adding extra carriages for the summer months. It could also be used in route planning and locomotive design.

4.6 Case Study: Accurate Buckling Load for a Cantilevered Shaft

In Section 4.3, an energy method was used to estimate the end load which just causes a cantilevered shaft to buckle. More accurate solutions to the same problem may now be obtained with the aid of the predictor–corrector method for solving differential equations. A 'shooting' method is needed to satisfy conditions at more than one value of the independent variable. The problem is not, however, quite of the boundary-value type discussed in Section 4.4.3, but is of the eigenvalue type. The value of end load is required which just causes nonzero but finite deflections of the shaft. As the situation is one of neutral equilibrium, the magnitudes of these finite deflections cannot be determined, although the deflected shape of the shaft can be found. An end load less than the critical value creates a condition of stable equilibrium and zero deflections, whereas a larger load causes instability and gross deflections. Other types of eigenvalue problems are discussed in Section 5.8.

Figure 4.5 shows the geometry and notation for the problem. If δ is the deflection in the y-direction of the free end of the shaft, the bending moment at a distance x from the fixed end is $F(\delta - y)$, where y is the local deflection. Therefore, the curvature is

$$\frac{d^2y}{dx^2} = \frac{F(\delta - y)}{EI} \qquad (4.33)$$

where E and I are as defined in Section 4.3. If F and δ are prescribed, the problem is of the initial-value type, with $y(0) = y'(0) = 0$. What is required, however, is the smallest value of F which gives a small but finite end deflection, δ. Higher values of F, associated with higher modes of buckling, do exist, but are not readily obtainable by the present numerical method.

If the value of δ in equation (4.33) is prescribed ($\delta \ll L$), then the end deflection, y_L, obtained by solving this equation is a function of the end load. The required critical value of the load is therefore the solution to the equation

$$y_L(F) = \delta \qquad (4.34)$$

In principle, the Newton–Raphson method described in Section 3.7.2 could be employed, the necessary derivatives being evaluated numerically. A simpler method is used here, however, to illustrate another method of solving nonlinear equations. Its main advantage is that it always locates a root if one exists, but its rate of convergence is relatively slow.

Figure 4.16 illustrates the technique, which may be described as a step-halving method. The curve, which passes through the origin, indicates the general form of the function $y_L(F)$. Provided an initial value of load, F_0, can be found such that the corresponding computed end deflection $y_L(F_0)$ is greater than δ, the method proceeds as follows. Initially it is known that the required critical load, F_C, lies between zero and F_0. This range is halved by choosing $F_1 = \frac{1}{2}F_0$, and the corresponding end deflection $y_L(F_1)$ is compared with δ. If, as in Fig.

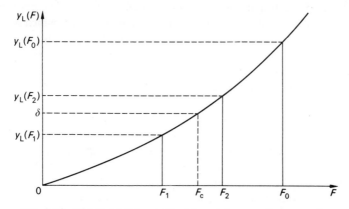

Fig. 4.16   The step-halving method for solving a nonlinear equation

4.16, δ is the larger of the two, the range has been narrowed to $F_1 < F_c < F_0$. The process is continued with $F_2 = \frac{1}{2}(F_0 + F_1)$, and $y_L(F_2) > \delta$ therefore $F_1 < F_c < F_2$, and so on until the calculated end deflection agrees with δ to within an acceptable tolerance.

4.6.1 Problem specification The problem specified in Section 4.3.1 is to be solved more accurately with the aid of the Adams–Bashforth predictor–corrector method. Solutions are to be obtained using the shooting method and step-halving technique described above, and are to be compared with those obtained by the energy method.

4.6.2 Solution Figures 4.17 to 4.19 show a main program and two subprograms for solving this problem. The predictor–corrector subprogram PRECOR shown in Fig. 4.11 is also used. The main program, Fig. 4.17, controls the input and output of data, and applies the step-halving technique to solve equation (4.34). The end deflection for a given end load is determined by the FUNCTION subprogram named YEND, shown in Fig. 4.18. This subprogram serves to control the solution of equation (4.33) by the predictor–corrector method. It therefore has many similarities to the main program of the last case study, Fig. 4.10. The SUBROUTINE subprogram named BFNS, Fig. 4.19, defines the derivatives required for the predictor–corrector process and is therefore similar in form to FNS, Fig. 4.12.

In Fig. 4.17, the variables D1, D2, L, E and NSTEP are defined as in Section 4.3.2, while DELTA is used to store the prescribed end deflection, δ. NITER and CYCLES store the total number of corrector iterations, and the mean number per step respectively, the values of which are returned from the deflection calculation in YEND. All of these variables are located in the COMMON block named CBUCK, for use in other subprograms. After writing out a program title and heading for the table of results, the main program reads in the relevant data, including an estimate of the buckling load, which is stored in the variable FO. A zero or negative value of NSTEP terminates execution, for example, by means of a blank card. The value which is prescribed for δ is arbitrary, but is small compared to the length of the shaft.

The choice of the value of FO read in is not crucial, because the first action of the step-halving method of solution is to ensure that it is greater than the critical value, if necessary by repeated doubling. In the unlikely event of the end deflection associated with FO not exceeding δ after the force is doubled 500 times, execution is terminated. This is only likely to happen if the input value is zero or negative. Note the concise way of calling the FUNCTION subprogram YEND, by simply using its name in a statement and inserting the appropriate argument. The step-halving procedure described above and illustrated in Fig. 4.16 is then applied. Upper and lower limits for the buckling load, FMAX and

```
C   PROGRAM TO CALCULATE SHAFT BUCKLING FORCE ACCURATELY.
C
        COMMON /CBUCK/ D1,D2,L,E,NSTEP,DELTA,NITER,CYCLES
        REAL  L
        WRITE(6,61)
61      FORMAT(45H1      SHAFT BUCKLING FORCE BY SHOOTING METHOD //
   1 70H      D1        D2      L       E        NSTEP     F       ITER NITE
   2R CYCLES)
1       READ(5,51) D1,D2,L,E,NSTEP,F0
51      FORMAT(3F10.0,E15.5,I5,E15.5)
        IF(NSTEP.GT.0) GO TO 2
        STOP
C
C   SET ARBITRARY END DEFLECTION.
2       DELTA=0.01*L
C
C   ENSURE THAT THE INITIAL GUESS FOR THE END LOAD IS TOO HIGH.
        DO 3 IF=1,500
        IF(YEND(F0).GT.DELTA) GO TO 4
3       F0=F0*2.
        WRITE(6,62)
62      FORMAT(32H REQUIRED INITIAL F NOT OBTAINED)
        STOP
C
C   STEP-HALVING METHOD OF SOLUTION.
4       FMAX=F0
        FMIN=0.
C
C   SET UP THE ITERATION LOOP.
        TOLER=1.E-6
        DO 5 ITER=1,500
        F=0.5*(FMAX+FMIN)
        YENDF=YEND(F)
        ERROR=ABS(YENDF/DELTA-1.)
C
C   TEST FOR CONVERGENCE.
        IF(ERROR.LT.TOLER) GO TO 6
        IF(YENDF.GT.DELTA) FMAX=F
5       IF(YENDF.LE.DELTA) FMIN=F
C
C   NORMAL EXIT FROM ITERATION LOOP INDICATES FAILURE TO CONVERGE.
        WRITE(6,63)
63      FORMAT(15H NO CONVERGENCE)
        STOP
C
C   OUTPUT THE RESULTS.
6       WRITE(6,64) D1,D2,L,E,NSTEP,F,ITER,NITER,CYCLES
64      FORMAT(1X,3F7.2,E12.4,I6,E12.4,I5,I6,F7.2)
        GO TO 1
        END
```

Fig. 4.17   Main program for the accurate determination of shaft buckling load

FMIN, are defined, and the range is repeatedly halved to give the current end load F, until the corresponding end deflection, YENDF, agrees with the prescribed value. A tolerance of 10^{-6} on the relative error is used. Finally, the input data and results are written out, before a new set of data is read.

In the subprogram YEND, Fig. 4.18, the arrays Y and F (not to be confused with the end load in the main program) are employed to store the functions and derivatives used in the predictor-corrector process. As in equations (4.18) to (4.20), the second-order equation (4.33) is reduced to two first-order equations, (4.35) and (4.36) where y_1 is the deflection and y_2 its slope. The current value of the end load, entered via the argument FARG, is assigned to the variable

FORCE which is located in its own COMMON block named CBFNS. This allows the value of the end load to be made available to the subprogram BFNS. An alternative form of coding would be to make YEND a subprogram without arguments and to store the end load in a variable located in block CBUCK.

$$\frac{dy_1}{dx} = f_1 = y_2 \tag{4.35}$$

$$\frac{dy_2}{dx} = f_2 = \frac{F(\delta - y_1)}{EI} \tag{4.36}$$

After checking the validity of NSTEP, the remainder of subprogram YEND is very similar in form to Fig. 4.10, which has been discussed in detail. Note that the FUNCTION name appears on the left-hand side of an arithmetic statement in order to store the computed end deflection. Subprogram BFNS, Fig. 4.19, is very similar in form to Fig. 4.12 and is used to define the two derivatives according to equations (4.35) and (4.36). The required second moment of area of the shaft is defined with the aid of equations (4.13) and (4.14). Note that a value for π is stored in the variable PI by means of a DATA statement when the

```
      FUNCTION   YEND(FARG)
C
C  FUNCTION SUBPROGRAM FOR FINDING THE ACTUAL SHAFT END DEFLECTION FOR
C  A GIVEN LOAD.
C
      DIMENSION  Y(600,2),F(600,2)
      COMMON /CBUCK/ D1,D2,L,E,NSTEP,DELTA,NITER,CYCLES   /CBFNS/ FORCE
      EXTERNAL   BFNS
      REAL   L
      FORCE=FARG
C
C  TEST VALIDITY OF SUPPLIED NUMBER OF STEPS.
      IF(NSTEP.LT.600) GO TO 1
      WRITE(6,61) NSTEP
   61 FORMAT(23H STOP IN YEND - NSTEP =,I5)
      STOP
C
C  SET INITIAL CONDITIONS.
    1 Y(1,1)=0.
      Y(1,2)=0.
      CALL  BFNS(Y,F,0.,1)
      NITER=0
      JMAX=NSTEP+1
      H=L/FLOAT(NSTEP)
C
C  APPLY THE PREDICTOR-CORRECTOR METHOD.
      DO 2 J=2,JMAX
      X=H*FLOAT(J-1)
      CALL  PRECOR(Y,F,X,2,600,J,H,ITER,BFNS)
C
C  ACCUMULATE TOTAL NUMBER OF CORRECTOR ITERATIONS.
    2 NITER=NITER+ITER
C
C  STORE THE REQUIRED END DEFLECTION AND THE AVERAGE NUMBER OF CORRECTOR
C  ITERATIONS.
      YEND=Y(JMAX,1)
      CYCLES=FLOAT(NITER)/FLOAT(NSTEP)
      RETURN
      END
```

Fig. 4.18   Subprogram for computing shaft end deflection

```
      SUBROUTINE  BFNS(Y,F,X,JA)
C
C  SUBROUTINE FOR DEFINING THE FUNCTIONAL FORMS OF THE DERIVATIVES.
C
      DIMENSION  Y(600,2),F(600,2)
      COMMON /CBUCK/ D1,D2,L,E,NSTEP,DELTA,NITER,CYCLES    /CBFNS/ FORCE
      REAL  L,I
      DATA  PI / 3.14159265 /
      J=JA
      D=D1+X/L*(D2-D1)
      I=PI*D**4/64.
      F(J,1)=Y(J,2)
      F(J,2)=FORCE*(DELTA-Y(J,1))/(E*I)
      RETURN
      END
```

Fig. 4.19   Subprogram for defining derivatives

program is compiled. A trigonometrical expression such as that used in Fig. 4.6 should not be employed here, as it would be re-evaluated every time BFNS is executed.

4.6.3 Results and discussion   Figure 4.20 shows the results obtained for this problem. Buckling loads are printed out under the heading 'F', while the values under 'ITER' are the numbers of iterations required in the step-halving solution procedure. The total number of corrector iterations and the mean number per step for the last application of the predictor–corrector are printed under the headings 'NITER' and 'CYCLES' respectively.

The first eight results are for the case of the parallel shaft, for which the exact solution is $F = 0.6449 \times 10^7$ N to four significant figures. The only parameter varied in these results is the number of steps along the shaft, which ranges from 2 to 256. To obtain four-figure accuracy, 128 steps are needed, and an average of approximately two corrector iterations per step are required. In Section 4.4.2, the latter condition was suggested as a criterion for choosing the step length. Note that further doubling of the number of steps results in comparatively small increases in the total number of corrector iterations. If greater accuracy is required than is available when using 128 steps, however, it would probably be

SHAFT BUCKLING FORCE BY SHOOTING METHOD

D1	D2	L	E	NSTEP	F	ITER	NITER	CYCLES
0.20	0.20	2.50	0.2080E+12	2	0.7175E+07	18	32	16.00
0.20	0.20	2.50	0.2080E+12	4	0.6619E+07	18	35	8.75
0.20	0.20	2.50	0.2080E+12	8	0.6491E+07	18	44	5.50
0.20	0.20	2.50	0.2080E+12	16	0.6460E+07	19	68	4.25
0.20	0.20	2.50	0.2080E+12	32	0.6452E+07	19	97	3.03
0.20	0.20	2.50	0.2080E+12	64	0.6450E+07	19	146	2.28
0.20	0.20	2.50	0.2080E+12	128	0.6449E+07	18	262	2.05
0.20	0.20	2.50	0.2080E+12	256	0.6449E+07	20	315	1.23
0.20	0.15	2.50	0.2080E+12	128	0.4545E+07	18	261	2.04
0.20	0.10	2.50	0.2080E+12	128	0.2690E+07	20	268	2.09

Fig. 4.20   Accurate buckling loads

Table 4.2
Comparison of buckling loads

D_1/D_2	Buckling loads (N x 10^7)		Difference (per cent)
	Accurate	Approximate	
1·0	0·6449	0·6449	0
0·75	0·4545	0·4821	6
0·5	0·2690	0·3671	36

better to employ a more accurate predictor–corrector method. The last two results are for the tapered shafts, using 128 steps in each case.

It is interesting to compare the present results with those shown in Fig. 4.7 for the energy method. Table 4.2 lists the predicted buckling loads and the percentage differences by which the approximate values exceed the corresponding accurate ones. As is to be expected, these differences increase as the true deflected shape of the shaft diverges from the shape assumed in the energy method. Note that this method always gives an upper bound for the buckling load.

5 Simultaneous Linear Equations

Many engineering problems involve finding the solutions to sets of simultaneous linear algebraic equations. These arise either directly from the mathematical model used to describe the problem, or as a result of applying a particular method of numerical analysis to the mathematical equations. In this chapter the case studies are of the former type, whereas in Chapter 6 the solution of simultaneous equations arising from the numerical analysis of partial differential equations is considered. Chapter 7 provides examples of both types, arising in the fields of structural analysis and finite element techniques.

Attention is confined here to sets of n simultaneous equations in n unknowns, $x_1, x_2, \ldots, x_n$, which may be expressed in the form

$$a_{11}x_1 + a_{12}x_2 + \ldots + a_{1n}x_n = b_1$$
$$a_{21}x_1 + a_{22}x_2 + \ldots + a_{2n}x_n = b_2 \qquad (5.1)$$
$$\cdots\cdots\cdots\cdots\cdots\cdots\cdots\cdots\cdots$$
$$a_{n1}x_1 + a_{n2}x_2 + \ldots + a_{nn}x_n = b_n$$

where the coefficients a_{ij} and b_i are all known constants. In this suffix notation, i specifies the equation or *row* number, while j defines the *column* number (which involves the unknown x_j). If the b_i are all zero, the equations are said to be homogeneous, and special treatment is required to avoid the trivial result that all the x_j are zero. Such treatment is considered in Section 5.8. If at least one of the b_i is nonzero, the equations are said to be nonhomogeneous. Equations (5.1) may be expressed in matrix form as follows

$$\begin{bmatrix} a_{11}\ a_{12}\ \ldots\ a_{1n} \\ a_{21}\ a_{22}\ \ldots\ a_{2n} \\ \cdots\cdots\cdots\cdots \\ a_{n1}\ a_{n2}\ \ldots\ a_{nn} \end{bmatrix} \begin{bmatrix} x_1 \\ x_2 \\ . \\ x_n \end{bmatrix} = \begin{bmatrix} b_1 \\ b_2 \\ . \\ b_n \end{bmatrix} \qquad (5.2)$$

$$AX = B \qquad (5.3)$$

The column matrices X and B are often called vectors. The object of 'solving' these equations is to determine the vector X, which can be written as

$$X = A^{-1}B \tag{5.4}$$

where A^{-1} is the inverse of matrix A. Thus, all methods of solution are equivalent to finding this inverse.

There are two main types of method for solving simultaneous linear equations, namely direct and iterative. Within these categories, numerous comparatively similar methods have been developed, and are described in detail by, for example, Bickley and Thompson (1964), Conte (1965), Ralston (1965) or Williams (1972). Attention is concentrated here on one of the most commonly used methods of each type, although other methods are discussed.

5.1   Direct Methods of Solution

The basis of most direct methods of solving simultaneous linear equations is the systematic elimination of the unknowns. Before considering elimination methods, however, it is useful to examine a method involving determinants.

5.1.1 Cramer's rule   In order to illustrate the method it is convenient to consider a particular set of equations

$$\begin{aligned}
2x_1 + x_2 - x_3 &= 1 \\
x_1 + 3x_2 + 2x_3 &= 13 \\
x_1 - x_2 + 4x_3 &= 11
\end{aligned} \tag{5.5}$$

Cramer's rule states that the unknowns are given by the following ratios of determinants

$$x_1 = \frac{|C_1|}{|A|}, \quad x_2 = \frac{|C_2|}{|A|}, \quad x_3 = \frac{|C_3|}{|A|} \tag{5.6}$$

where $|A|$ is the determinant of the coefficient matrix A. C_1, C_2 and C_3 are matrices obtained by replacing the first, second and third columns of A respectively, by the vector B (in the notation of equation (5.3)). Thus

$$|A| = \begin{vmatrix} 2 & 1 & -1 \\ 1 & 3 & 2 \\ 1 & -1 & 4 \end{vmatrix} = 30, \quad |C_1| = \begin{vmatrix} 1 & 1 & -1 \\ 13 & 3 & 2 \\ 11 & -1 & 4 \end{vmatrix} = 30 \tag{5.7}$$

and similarly for $|C_2| = 60$, $|C_3| = 90$, giving $x_1 = 1$, $x_2 = 2$, $x_3 = 3$. While this method is convenient for a small number of equations, it is very inefficient for a large set, as discussed in Section 5.5.1.

Although Cramer's rule is inappropriate for solving large sets of equations, it does indicate when difficulties may be expected. Clearly, equations (5.6) are only meaningful if A is nonsingular, that is $|A| \neq 0$. Problems involving singular coefficient matrices are considered in Section 5.8. Difficulties may also be experienced when $|A|$ is small compared to the coefficients or *elements* of A, and the equations are said to be 'ill-conditioned'.

5.1.2 Gaussian elimination

Elimination methods involve the successive elimination of the unknowns by algebraic manipulation. Although a number of different methods have been evolved, they are fundamentally very similar to the gaussian elimination method described here. The straightforward elimination scheme is considered first, without refinements such as *partial pivoting*.

In equations (5.1), the first equation can be used to eliminate x_1 from the remaining $n-1$ equations. The modified second equation is then used to eliminate x_2 from the remaining $n-2$ equations, and so on until the last equation contains only x_n. Thus x_n may be found, followed by all the other unknowns, by back substitution. Consider, for example, equations (5.5). The first is divided through by 2 and subtracted from the second and third equations

$$
\begin{aligned}
2x_1 + \quad x_2 - \quad x_3 &= 1 \\
2 \cdot 5x_2 + 2 \cdot 5x_3 &= 12 \cdot 5 \\
-1 \cdot 5x_2 + 4 \cdot 5x_3 &= 10 \cdot 5
\end{aligned}
\tag{5.8}
$$

The second of these is multiplied through by $1 \cdot 5 \div 2 \cdot 5$ and added to the third equation

$$
\begin{aligned}
2x_1 + \quad x_2 - \quad x_3 &= 1 \\
2 \cdot 5x_2 + 2 \cdot 5x_3 &= 12 \cdot 5 \\
6x_3 &= 18
\end{aligned}
\tag{5.9}
$$

By back substitution $x_3 = 3$, $x_2 = 5 - x_3 = 2$, $x_1 = \frac{1}{2}(1 - x_2 + x_3) = 1$.

For the general equations (5.1), let the coefficients be given the notation $a_{ij}^{(1)}$, $b_i^{(1)}$. After the first elimination, the modified coefficients are

$$
a_{ij}^{(2)} = a_{ij}^{(1)} - \phi \, a_{1j}^{(1)}
$$

$$
b_i^{(2)} = b_i^{(1)} - \phi \, b_i^{(1)}
\tag{5.10}
$$

$$
\phi = a_{i1}^{(1)} / a_{11}^{(1)}, \ i = 2, 3, \ldots, n \text{ and } j = 1, 2, \ldots, n
$$

Similarly, after the kth elimination

$$a_{ij}^{(k+1)} = a_{ij}^{(k)} - \phi\, a_{kj}^{(k)}$$

$$b_i^{(k+1)} = b_i^{(k)} - \phi\, b_k^{(k)} \qquad (5.11)$$

$$\phi = a_{ik}^{(k)}/a_{kk}^{(k)},\ i = k+1, k+2, \ldots, n \text{ and } j = k, k+1, \ldots, n$$

Note that the vector B is treated just like a column of A, and advantage can be taken of this fact to simplify the programming of the process. The final set of equations is

$$a_{11}^{(1)}x_1 + a_{12}^{(1)}x_2 + \ldots + a_{1n}^{(1)}x_n = b_1^{(1)}$$

$$a_{22}^{(2)}x_2 + \ldots + a_{2n}^{(2)}x_n = b_2^{(2)} \qquad (5.12)$$

$$\cdots\cdots\cdots\cdots\cdots\cdots\cdots\cdots\cdots$$

$$a_{nn}^{(n)}x_n = b_n^{(n)}$$

Expressed in matrix terminology, this elimination process triangularises A. All the elements below the *diagonal* (that is, a_{ij} with $i > j$) are reduced to zero, leaving an upper-triangular matrix. The final solution by back substitution is equivalent to inverting this triangular matrix. The unknowns are obtained in reverse order

$$x_n = b_n^{(n)}/a_{nn}^{(n)} \qquad (5.13)$$

$$x_i = \left(b_i^{(i)} - \sum_{j=i+1}^{n} a_{ij}^{(i)}x_j\right)/a_{ii}^{(i)} \qquad (5.14)$$

$$i = n-1, n-2, \ldots, 1$$

As the elimination process does not affect the value of $|A|$, it may be found from the triangularised matrix as

$$|A| = a_{11}^{(1)}a_{22}^{(2)}a_{33}^{(3)} \ldots\ldots\ldots a_{nn}^{(n)} \qquad (5.15)$$

For example, from equations (5.9), $|A| = 2 \times 2 \cdot 5 \times 6 = 30$, which agrees with equations (5.7).

A great many arithmetic operations are involved in solving large sets of equations by elimination. Any errors introduced, such as roundoff errors, tend to be magnified and may become unacceptably large. Equations (5.11) show that the elimination process involves many multiplications by the factors ϕ. In order to minimise the effect of any errors in the coefficients $a_{kj}^{(k)}$ and $b_k^{(k)}$, these factors should be as small as possible, and certainly less than one. Thus, the pivotal element $a_{kk}^{(k)}$ should be the largest element in the 'leading' column of the remaining submatrix

$$\left|a_{kk}^{(k)}\right| > \left|a_{ik}^{(k)}\right| \quad i = k+1, k+2, \ldots, n \qquad (5.16)$$

This condition also serves to avoid division by zero in equations (5.11), and can be achieved by a technique known as partial pivoting. Immediately before each elimination, the leading column is searched for the largest element. By interchanging equations, this can be made the pivotal element to satisfy equation (5.16). This process can be illustrated with the following equations

$$
\begin{aligned}
x_1 - 2x_2 + 3x_3 &= 2 \\
3x_1 + x_2 - 2x_3 &= 9 \\
4x_1 - 2x_2 + x_3 &= 9
\end{aligned}
\qquad (5.17)
$$

The largest coefficient of x_1 occurs in the third equation, which is interchanged with the first before the first elimination is performed to give

$$
\begin{aligned}
4x_1 - 2x_2 + x_3 &= 9 \\
2 \cdot 5x_2 - 2 \cdot 75x_3 &= 2 \cdot 25 \\
-1 \cdot 5x_2 + 2 \cdot 75x_3 &= -0 \cdot 25
\end{aligned}
\qquad (5.18)
$$

As the coefficient of x_2 in the second of these equations is larger in magnitude than the corresponding coefficient in the third equation, no interchange is required before the second elimination. The final solution is obtained as $x_1 = 3$, $x_2 = 2, x_3 = 1$.

The idea of partial pivoting can be extended to searching the whole of the remaining submatrix for the largest element. Such complete pivoting involves interchanging both rows and columns, and is more difficult to program. Since it offers only slight improvement in accuracy over partial pivoting, it is rarely used.

Another possible refinement is to scale the equations to make their coefficients similar in magnitude. One way to do this is to normalise each equation so that the largest element in each row of A is one. Scaling can be important when corresponding coefficients in different equations differ by several orders of magnitude.

5.1.3 Other direct methods Apart from Cramer's rule and elimination techniques, there are other direct methods for solving simultaneous linear equations. One of the most important types is that of matrix factorisation. The usual approach is to factorise A into a product of a lower-triangular and an upper-triangular matrix. As with elimination methods, a triangular set of equations is readily solved. Matrix factorisation methods are fundamentally very similar to elimination techniques, and normally involve a similar number of arithmetic operations. They are particularly useful for symmetric matrices (with $a_{ij} = a_{ji}$ for all i, j) where advantage can be taken of the symmetry to reduce the amount of arithmetic and storage required.

5.1.4 Special forms of equations The most important special forms of equations are those where many of the coefficients are zero. In Chapter 6, for example, it is shown that such *sparse* equations often occur in numerical methods for solving partial differential equations. Direct methods of solution which require all the coefficients to be stored are wasteful of core storage for such equations. As discussed in Section 5.5, iterative methods of solution are often preferred for sparse equations.

In the particular case of sparse equations with a tridiagonal coefficient matrix, however, the solutions can be obtained by a direct method with a minimum of computing time and storage. If A is tridiagonal, all elements other than those on the diagonal and immediately adjacent to it are zero. Thus

$$A = \begin{bmatrix} a_{11} & a_{12} & & & & \\ a_{21} & a_{22} & a_{23} & & & \\ & a_{32} & a_{33} & a_{34} & & \\ & & \cdot & \cdot & \cdot & \\ & & & \cdot & \cdot & \cdot \\ & & & & a_{nn-1} & a_{nn} \end{bmatrix} \qquad (5.19)$$

Such matrices occur in boundary-value problems for ordinary differential equations, and the method of solution is considered in Section 6.3.2.

5.2 Iterative Methods of Solution

The basis of all iterative methods for solving simultaneous linear equations is to first guess values for the unknowns, and to devise a scheme for successively improving these values. Such a scheme involves expressing each unknown as a function of the others, as follows

$$\begin{aligned} x_1 &= f_1(x_2, x_3, \ldots x_n) \\ x_2 &= f_2(x_1, x_3, \ldots x_n) \\ &\cdots\cdots\cdots\cdots\cdots \\ x_n &= f_n(x_1, x_2, \ldots x_{n-1}) \end{aligned} \qquad (5.20)$$

There are similarities with the functional iteration procedure for nonlinear equations, described in Section 3.7.1. Here, the functions are linear, and the subject of each equation appears only on the left-hand side. Estimates of the unknowns are substituted into the right hand sides to give new values which, if the process is convergent, provide successively better approximations to the required solutions.

Consider, for example, equations (5.5), which may be rearranged as follows

$$x_1 = \tfrac{1}{2}(1 - x_2 + x_3)$$
$$x_2 = \tfrac{1}{3}(13 - x_1 - 2x_3) \tag{5.21}$$
$$x_3 = \tfrac{1}{4}(11 - x_1 + x_2)$$

If initial guesses of $x_1 = x_2 = x_3 = 0$ are substituted into these equations, the new values are $x_1 = 0 \cdot 5, x_2 = 4 \cdot 33$ and $x_3 = 3 \cdot 75$. After the second iteration the values are $x_1 = -0 \cdot 29$, $x_2 = 2 \cdot 33$, $x_3 = 3 \cdot 71$, and eventually the correct results $x_1 = 1$, $x_2 = 2$, $x_3 = 3$ are obtained. This is known as Jacobi's method, but is rarely used because there are various ways to improve the rate of convergence.

5.2.1 Gauss–Seidel method   In the above iteration scheme, the values of the unknowns substituted into the equations are those obtained from the previous cycle of iteration. One possible improvement is to always use the latest values. For example, using the same initial guesses, the first application of equations (5.21) yields

$$x_1 = \tfrac{1}{2}(1 - 0 + 0) = 0 \cdot 5$$
$$x_2 = \tfrac{1}{3}(13 - 0 \cdot 5 - 0) = 4 \cdot 167 \tag{5.22}$$
$$x_3 = \tfrac{1}{4}(11 - 0 \cdot 5 + 4 \cdot 167) = 3 \cdot 667$$

which are rather better than the corresponding values obtained by Jacobi's method. The present technique, which is known as the Gauss–Seidel method, generally gives a faster rate of convergence. For equations (5.21), the solutions converge after thirteen cycles to a tolerance of 10^{-6} as defined in equation (5.25) below.

The Gauss–Seidel process may be generalised for equations (5.1) as follows

$$x_i^{(m)} = \frac{1}{a_{ii}} \left(b_i - \sum_{j=1}^{i-1} a_{ij} x_j^{(m)} - \sum_{j=i+1}^{n} a_{ij} x_j^{(m-1)} \right) \tag{5.23}$$

where the superscripts denote iteration numbers. Clearly, it is essential for all the diagonal elements to be nonzero, and for the process to be convergent even more stringent conditions should be satisfied.

5.2.2 Convergence of iterative methods Following the principles of testing for convergence outlined in Section 3.1, it is appropriate to compare the changes in the x_i between successive iterations with their current values. A possible convergence criterion is

$$e_r = \sum_{i=1}^{n} \left| \frac{\delta x_i}{x_i^{(m)}} \right| < \alpha \tag{5.24}$$

where $\delta x_i = x_i^{(m)} - x_i^{(m-1)}$ and α is a suitably small tolerance. This is a poor choice of criterion, however, as one or more of the x_i may be very small or zero. A much better criterion, which overcomes this difficulty, is

$$e_r = \sum_{i=1}^{n} |\delta x_i| \Big/ \sum_{i=1}^{n} |x_i^{(m)}| < \alpha \tag{5.25}$$

Divergence of the process, where the x_i tend to infinity, is more difficult to define. A useful test is for $e_r > 1$, particularly after the first one or two iterations, although this will not detect slow divergence.

The choice of starting values for the unknowns does not normally affect whether the Gauss–Seidel process converges, and often has comparatively little effect on the number of iterations required. It is possible to predict whether convergence is likely to be achieved with a particular set of equations. Varga (1962) has stated the sufficient condition for convergence as that of 'diagonal dominance' of the coefficient matrix A. If A is diagonally dominant, the magnitudes of its elements are such that

$$|a_{ii}| \geqslant \sum_{\substack{j=1 \\ j \neq i}}^{n} |a_{ij}| \tag{5.26}$$

and the inequality is satisfied for at least one row. While diagonal dominance is sufficient to ensure convergence, it may not be necessary, provided these conditions are only mildly contravened. Clearly, sets of equations should be arranged to promote diagonal dominance before the Gauss–Seidel method is applied. Equations (5.5), which were used to illustrate iterative methods of solution, do satisfy equations (5.26). Equations (5.17), on the other hand, cannot be rearranged to satisfy diagonal dominance, and are unsuitable for iterative methods of solution.

5.2.3 Over-relaxation It is often possible to improve the rate of convergence of the Gauss–Seidel method by a technique which is generally known as over-relaxation. Equation (5.23) provides new estimates, $x_i^{(m)}$, which, provided the process is convergent, are closer to the required solutions than the $x_i^{(m-1)}$ Over-relaxation applies a limited amount of extrapolation from these two sets of estimates towards the final solutions. Thus, if $\tilde{x}_i^{(m)}$ are the values obtained from equations (5.23), the extrapolated values after the mth iteration are

$$x_i^{(m)} = x_i^{(m-1)} + \omega \left(\tilde{x}_i^{(m)} - x_i^{(m-1)} \right) \qquad (5.27)$$

where ω is an over-relaxation factor, which is the same for all the equations. For a particular set of linear equations, there is an optimum value of ω, normally in the range $1 < \omega < 2$, although sometimes under-relaxation ($\omega < 1$) is beneficial. The purpose of over-relaxation is to accelerate convergence, rather than to promote convergence in an otherwise divergent iteration scheme. The use of too large a value of ω can cause divergence.

For programming purposes, it is convenient to rewrite equations (5.27) and (5.23) with the aid of the changes in the unknowns, δx_i, introduced in equation (5.24). Thus

$$x_i^{(m)} = x_i^{(m-1)} + \omega \, \delta x_i$$

$$\delta x_i = \frac{1}{a_{ii}} \left(b_i - \sum_{j=1}^{n} a_{ij} x_j \right)$$

(5.28)

where the latest values of the unknowns are used in the summations.

Unfortunately, it is not a simple matter to predict the optimum value of ω (see Varga (1962), and Isaacson and Keller (1966)), and the usual approach is an empirical one. The required value can be found as that which gives either convergence to a particular tolerance with the minimum number of iterations, or the minimum error after a certain number of iterations. The optimum over-relaxation factor is determined by the number of equations and the nature of the coefficient matrix, A.

The use of over-relaxation is particularly beneficial for the type of sparse equations that arise in the numerical solution of boundary-value problems for partial differential equations. Section 6.6 provides an example of the application of over-relaxation, while in the case study described in Section 5.4 under-relaxation is employed.

5.3 Case Study: Currents in an Electrical Network

Simultaneous linear equations often occur in electrical network problems. For example, the unbalanced Wheatstone bridge circuit shown in Fig. 5.1 gives rise to a set of six equations which is comparatively tedious to solve by hand. The potential difference and resistances are shown on the diagram, as are the assumed directions of the six unknown direct currents. Application of Kirchhoff's laws for the conservation of current at the four nodes of the network, and the

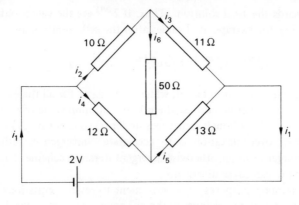

Fig. 5.1   An unbalanced Wheatstone bridge circuit

summation of potential differences round two closed loops, yields the following
set of equations for the currents in amperes.

$$
\begin{bmatrix}
1 & -1 & 0 & -1 & 0 & 0 \\
0 & 1 & -1 & 0 & 0 & -1 \\
0 & 0 & 0 & 1 & -1 & 1 \\
-1 & 0 & 1 & 0 & 1 & 0 \\
0 & 0 & 0 & 12 & 13 & 0 \\
0 & 10 & 11 & 0 & 0 & 0
\end{bmatrix}
\begin{bmatrix}
i_1 \\ i_2 \\ i_3 \\ i_4 \\ i_5 \\ i_6
\end{bmatrix}
=
\begin{bmatrix}
0 \\ 0 \\ 0 \\ 0 \\ 2 \\ 2
\end{bmatrix}
\qquad (5.29)
$$

5.3.1 Problem specification   The electrical currents in the network shown in
Fig. 5.1 are to be found by the direct gaussian elimination method for solving
simultaneous linear equations. If possible, the same results are to be obtained by
the Gauss–Seidel iterative method.

5.3.2 Solution   Figures 5.2 and 5.3 show a main program and gaussian
elimination subprogram named ELIMIN, for the direct solution of simultaneous
linear equations. A flow chart for ELIMIN is shown in Fig. 5.4.

In the main program, the DIMENSION statement allows for up to twenty
simultaneous equations to be solved. The array A is used to store the elements of
both the matrix A and the vector B defined in equation (5.3). As shown in
equations (5.11), B is treated just like a column of A during the elimination
process. Consequently, the number of columns in array A exceeds the number of
rows by one, and the values of the elements b_i are stored in the variables

A(I,JMAX). The variable I serves as the row counter, while JMAX stores the maximum value of the column counter, J, as one more than the number of equations. The subscripted variable X is used to store the required solutions.

The number of equations to be solved is read into the variable NEQN, and a test is applied both to prevent array dimensions being exceeded, and to provide a means of terminating execution with a blank card. Then the elements of the

```
C   SOLUTION OF SIMULTANEOUS LINEAR EQUATIONS BY GAUSSIAN ELIMINATION.
C
        DIMENSION  A(20,21),X(20)
    1   READ(5,51) NEQN
   51   FORMAT(I5)
C
C   TEST VALIDITY OF THE NUMBER OF EQUATIONS.
        IF(NEQN.LT.1.OR.NEQN.GT.20) STOP
        JMAX=NEQN+1
C
C   INPUT THE ELEMENTS OF THE EXTENDED COEFFICIENT MATRIX.
        READ(5,52) ((A(I,J),J=1,JMAX),I=1,NEQN)
   52   FORMAT(8F10.0)
        WRITE(6,61)
   61   FORMAT(66H1SOLUTION OF SIMULTANEOUS LINEAR EQUATIONS BY GAUSSIAN E
       1LIMINATION)
        DO 2 I=1,NEQN
    2   WRITE(6,62) (A(I,J),J=1,JMAX)
   62   FORMAT(1H0,12F10.5)
C
C   APPLY GAUSSIAN ELIMINATION.
        CALL  ELIMIN(A,X,NEQN,20,21,DETA,RATIO)
        IF(DETA.EQ.0.) GO TO 3
C
C   OUTPUT THE RESULTS.
        WRITE(6,63)
   63   FORMAT(34HOTHE SOLUTION VECTOR IS AS FOLLOWS)
        WRITE(6,62) (X(I),I=1,NEQN)
        WRITE(6,64) DETA,RATIO
   64   FORMAT(14HODETERMINANT =,E12.4,5X,8H RATIO =,E12.4)
        GO TO 1
    3   WRITE(6,65)
   65   FORMAT(51HOTHE EQUATIONS ARE VERY ILL-CONDITIONED OR SINGULAR)
        GO TO 1
        END
```

Fig. 5.2   Main program for solving simultaneous linear equations by gaussian elimination

extended coefficient matrix are read in, and immediately written out following a program title. Control of execution is then transferred to subprogram ELIMIN, and the solutions are returned to the main program in the array X. The arguments DETA and RATIO return the value of $|A|$ and a relative magnitude for this determinant, as described below. If the value of DETA is exactly zero, a condition imposed in ELIMIN when the equations are very ill conditioned, an appropriate message is written out in place of the solutions.

5.3.3 The gaussian elimination subprogram   The arguments of the subprogram ELIMIN shown in Fig. 5.3 include the array of coefficients, A, and the solution vector, X. It should be noted that the supplied elements of A are destroyed

```
      SUBROUTINE  ELIMIN(A,X,MEQN,NROW,NCOL,DET,RATIO)
C
C  SUBROUTINE FOR SOLVING SIMULTANEOUS LINEAR EQUATIONS BY GAUSSIAN
C  ELIMINATION WITH PARTIAL PIVOTING.
C
      DIMENSION  A(NROW,NCOL),X(NROW)
      NEQN=MEQN
      IF(NEQN.LE.NROW.AND.NEQN.LE.NCOL-1) GO TO 1
      WRITE(6,61)
  61  FORMAT(33HOSTOP - DIMENSION ERROR IN ELIMIN)
      STOP
C
C  FIND MEAN COEFFICIENT MAGNITUDE.
  1   AMEAN=0.
      DO 2 I=1,NEQN
      DO 2 J=1,NEQN
  2   AMEAN=AMEAN+ABS(A(I,J))
      AMEAN=AMEAN/FLOAT(NEQN*NEQN)
C
C  COMMENCE ELIMINATION PROCESS.
      JMAX=NEQN+1
      NEQNM1=NEQN-1
      DO 6 IEQN=1,NEQNM1
C
C  SEARCH LEADING COLUMN OF THE COEFFICIENT MATRIX FROM THE DIAGONAL
C  DOWNWARDS FOR THE LARGEST ELEMENT AND MAKE THIS THE PIVOTAL ELEMENT.
      IMIN=IEQN+1
      IMAX=IEQN
      DO 3 I=IMIN,NEQN
  3   IF(ABS(A(I,IEQN)).GT.ABS(A(IMAX,IEQN))) IMAX=I
      IF(IMAX.EQ.IEQN) GO TO 5
      DO 4 J=IEQN,JMAX
      AA=A(IEQN,J)
      A(IEQN,J)=A(IMAX,J)
  4   A(IMAX,J)=AA
C
C  ELIMINATE X(IEQN) FROM EQUATIONS (IEQN+1) TO NEQN, FIRST TESTING FOR
C  NONZERO PIVOTAL ELEMENT.
  5   IF(ABS(A(IEQN,IEQN)/AMEAN).LT.1.E-8) GO TO 10
      DO 6 I=IMIN,NEQN
      FACT=A(I,IEQN)/A(IEQN,IEQN)
      DO 6 J=IMIN,JMAX
  6   A(I,J)=A(I,J)-FACT*A(IEQN,J)
C
C  SOLVE THE UPPER-TRIANGULAR SET OF EQUATIONS BY BACK SUBSTITUTION.
      IF(ABS(A(NEQN,NEQN)/AMEAN).LT.1.E-8) GO TO 10
      X(NEQN)=A(NEQN,JMAX)/A(NEQN,NEQN)
      DO 8 L=2,NEQN
      I=NEQN+1-L
      SUM=A(I,JMAX)
      IP1=I+1
      DO 7 J=IP1,NEQN
  7   SUM=SUM-A(I,J)*X(J)
  8   X(I)=SUM/A(I,I)
C
C  EVALUATE DETERMINANT OF COEFFICIENT MATRIX AND COMPARE WITH
C  ORIGINAL COEFFICIENTS.
      DETA=1.
      DO 9 I=1,NEQN
  9   DETA=DETA*A(I,I)
      DET=DETA
      RATIO=DETA/AMEAN**NEQN
      RETURN
  10  DET=0.
      RETURN
      END
```

Fig. 5.3   Subprogram for gaussian elimination

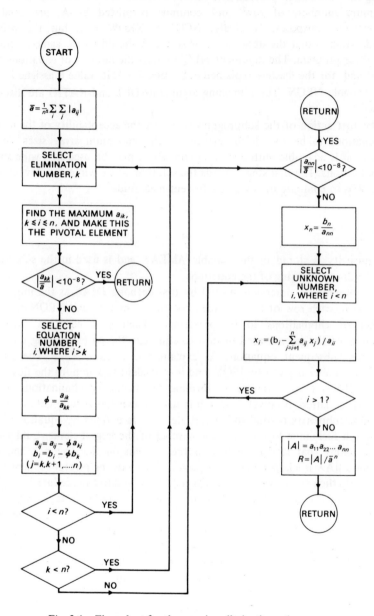

Fig. 5.4   Flow chart for the gaussian elimination subprogram

during the elimination process. The variables NROW and NCOL, which enter the maximum numbers of rows and columns permitted in A, are used for dimensioning purposes. Normally, NCOL = NROW + 1, but the primary consideration is that the dimensions of A and X should be the same as those in the calling program. The argument MEQN enters the number of equations to be solved and, for the reasons explained in Section 4.5.4, its value is assigned to the local variable NEQN. The remaining arguments DET and RATIO are discussed below.

The first action of the subprogram is to test the acceptability of the number of equations to be solved, in relation to the maximum array sizes. Such a precaution is advisable with a subprogram which may be used in a wide variety of applications. The first step towards the solution, as shown in the flow chart of Fig. 5.4, is to compute the mean coefficient magnitude

$$\bar{a} = \frac{1}{n^2} \sum_{i=1}^{n} \sum_{j=1}^{n} |a_{ij}| \qquad (5.30)$$

This quantity is stored in the variable AMEAN and is used in the subsequent tests for ill-conditioning of the equations.

The elimination process is started by first defining JMAX as the number of elements in each row of the extended coefficient matrix, and NEQNM1 as the number of eliminations to be performed. Then each equation, with the exception of the last, is used in turn to eliminate the corresponding unknown from the subsequent equations, as explained in Section 5.1.2. The current equation number is given by IEQN, and is equivalent to k in both the flow chart and equations (5.11). Before performing the necessary eliminations with a particular equation, however, a search is made down the leading column of the remaining submatrix to find the largest element, as defined by equation (5.16). The search technique locates the row number of the largest element, IMAX, by first assuming that it corresponds to the element on the diagonal, and only changing IMAX if a larger element is found. If the existing pivotal element is not the largest, then the rows are interchanged. An alternative procedure to actually interchanging rows is to store a record of the appropriate order of the rows.

Despite the search for the largest element, it is still possible for the resulting pivotal element to be extremely small or zero. This happens if the equations are very ill conditioned or singular, as explained in Section 5.6, and roundoff errors make it difficult to distinguish between these two states. The following test is made for the relative magnitude of the pivotal element

$$|a_{kk}/\bar{a}| < 10^{-8} \qquad (5.31)$$

and if this condition is satisfied at any stage of the elimination process, the problem is rejected. Rejection is indicated by setting the value of DET to zero, which may be detected by the calling program as described above. If the partial pivoting is successful, however, the eliminations defined by equations (5.11) are performed, with the variable FACT being used to store the values of the factor ϕ.

After testing the magnitude of the last diagonal element ($a_{nn}^{(n)}$ in equations (5.12)), the back substitutions defined in equations (5.13) and (5.14) are performed to find the required solutions. The variable SUM is used to accumulate the results of the summations indicated in equation (5.14). Note that the solution subscript, I (equivalent to i in equation (5.14)), cannot be used as the index of the outer DO loop controlling back substitution, because solutions must be obtained in reverse order.

Finally, the determinant of the coefficient matrix is obtained using equation (5.15). The local variable DETA is used to accumulate the required product, whose value is then assigned to the argument DET. Section 5.6 describes a test for ill-conditioning of the equations, which requires the following ratio

$$R = |A|/\bar{a}^n \tag{5.32}$$

whose value is stored in the argument RATIO.

5.3.4 Results Figure 5.5 shows the results of an attempt to solve equations (5.29) by gaussian elimination. The coefficients are printed out, but a very small or zero pivotal element is detected by the elimination subprogram, which stops the solution procedure. This is due to the selection of a singular set of equations, which was made deliberately to demonstrate the ability of the program to handle such equations. Examination of equations (5.29) shows that the sum of the first and fourth is identical to the sum of the second and third equations. If there are m nodes in an electrical network, Kirchhoff's law for current conservation should only be applied to m-1 of them if independent equations are to be obtained.

In order to solve the problem, one of the equations for current conservation, say the fourth, should be rejected in favour of one for potential difference which involves the 50 Ω resistor in Fig. 5.1. A suitable equation is

$$10i_2 + 50i_6 + 13i_5 = 2 \tag{5.33}$$

```
SOLUTION OF SIMULTANEOUS LINEAR EQUATIONS BY GAUSSIAN ELIMINATION
    1.00000  -1.00000   0.00000  -1.00000   0.00000   0.00000   0.00000
    0.00000   1.00000  -1.00000   0.00000   0.00000  -1.00000   0.00000
    0.00000   0.00000   0.00000   1.00000  -1.00000   1.00000   0.00000
   -1.00000   0.00000   1.00000   0.00000   1.00000   0.00000   0.00000
    0.00000   0.00000   0.00000  12.00000  13.00000   0.00000   2.00000
    0.00000  10.00000  11.00000   0.00000   0.00000   0.00000   2.00000
THE EQUATIONS ARE VERY ILL-CONDITIONED OR SINGULAR
```

Fig. 5.5   Results for gaussian elimination applied to equations (5.29)

When the revised coefficients are supplied to the program, the results shown in Fig. 5.6 are obtained. The printed solution vector displays the currents i_1 to i_6 from left to right. Since they are all positive, the directions assumed in Fig. 5.1 are correct. Note that i_6 is by far the smallest current.

By the criterion defined in Section 5.6, the equations are now well conditioned, because the printed value of RATIO is large compared to one. The program can be checked by substituting the solutions into the relevant equations.

```
SOLUTION OF SIMULTANEOUS LINEAR EQUATIONS BY GAUSSIAN ELIMINATION

   1.00000   -1.00000    0.00000   -1.00000    0.00000    0.00000    0.00000

   0.00000    1.00000   -1.00000    0.00000    0.00000   -1.00000    0.00000

   0.00000    0.00000    0.00000    1.00000   -1.00000    1.00000    0.00000

   0.00000   10.00000    0.00000    0.00000   13.00000   50.00000    2.00000

   0.00000    0.00000    0.00000   12.00000   13.00000    0.00000    2.00000

   0.00000   10.00000   11.00000    0.00000    0.00000    0.00000    2.00000

THE SOLUTION VECTOR IS AS FOLLOWS

   0.17524    0.09530    0.09518    0.07994    0.08006    0.00012

DETERMINANT = -0.3228E+05      RATIO = -0.1597E+02
```

Fig. 5.6   Currents in the Wheatstone bridge circuit

The present set of linear equations, in the order implied by the coefficient display in Fig. 5.6, are not diagonally dominant according to equation (5.26). Even if the order is rearranged, the diagonal dominance condition cannot be sufficiently satisfied for the Gauss–Seidel method of solution to be applicable.

5.4   Case Study: Stresses in Concentric Thick-Walled Cylinders

Figure 5.7 shows a cross-sectional view of two long concentric thick-walled cylinders. At ambient temperature and in the unstressed state, these cylinders have inner and outer radii of r_1 and r_2, and r_2 and r_3. An internal pressure of P above ambient is applied in such a way that no axial stresses are generated in the cylinders.

Using the Lamé equations (see, for example, Ford (1963)), the distributions of hoop and radial stresses may be expressed in dimensionless form for either cylinder as

$$\frac{\sigma_{\theta\theta}}{P} = Y + \frac{Z}{r^2}, \qquad \frac{\sigma_{rr}}{P} = Y - \frac{Z}{r^2} \qquad (5.34)$$

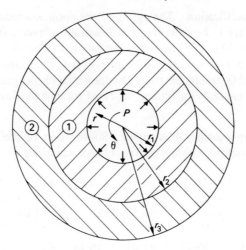

Fig. 5.7   Concentric thick-walled cylinders

where r and θ are the polar coordinates shown in Fig. 5.7. The parameters Y and Z are constants, whose values are determined by the boundary conditions for the particular cylinder. For the present problem, these conditions include: $\sigma_{rr} = -P$ at $r = r_1$, $\sigma_{rr} = 0$ at $r = r_3$, and the radial stresses are the same in both cylinders at $r = r_2$. Assuming the cylinders fit perfectly in the unstressed state, the hoop strains are also the same at $r = r_2$. Hoop strain is given by

$$e_{\theta\theta} = \frac{1}{E}(\sigma_{\theta\theta} - \nu\,\sigma_{rr}) = \frac{P}{E}\left(Y(1-\nu) + \frac{Z}{r^2}(1+\nu)\right) \qquad (5.35)$$

where E is Young's modulus and ν is Poisson's ratio. The following four equations are obtained

$$Y_1 - \frac{Z_1}{r_1^2} = -1$$

$$Y_1 - \frac{Z_1}{r_2^2} = Y_2 - \frac{Z_2}{r_2^2} \qquad (5.36)$$

$$Y_2 - \frac{Z_2}{r_3^2} = 0$$

$$\frac{1}{E_1}\left(Y_1(1-\nu_1) + \frac{Z_1}{r_2^2}(1+\nu_1)\right) = \frac{1}{E_2}\left(Y_2(1-\nu_2) + \frac{Z_2}{r_2^2}(1+\nu_2)\right)$$

The subscripts 1 and 2 refer to the inner and outer cylinders respectively.

5.4.1 Problem specification   The ratios of the hoop and radial stresses to the internal pressure are to be determined as functions of radius for the concentric cylinders shown in Fig. 5.7. The following data are to be used: $\nu_1 = 0.35$, $\nu_2 = 0.30$, $E_1/E_2 = 1.5$, $r_1 = 0.1$, $r_2 = 0.2$ and $r_3 = 0.4$, where the unit of radius is arbitrary. If possible, results are to be obtained by both the gaussian elimination and Gauss–Seidel methods for finding the Lamé parameters Y and Z.

5.4.2 Solution by gaussian elimination   Substitution of the above data into equations (5.36) yields the following set of linear equations

$$\begin{bmatrix} 1 & -100 & 0 & 0 \\ 1 & -25 & -1 & 25 \\ 0 & 0 & 1 & -6.25 \\ 0.65 & 33.75 & -1.05 & -48.75 \end{bmatrix} \begin{bmatrix} Y_1 \\ Z_1 \\ Y_2 \\ Z_2 \end{bmatrix} = \begin{bmatrix} -1 \\ 0 \\ 0 \\ 0 \end{bmatrix} \qquad (5.37)$$

Figure 5.8 shows the results obtained from the gaussian elimination program for this set of equations. The solutions are $Y_1 = 0.12500$, $Z_1 = 0.01125$, $Y_2 = 0.05208$ and $Z_2 = 0.00833$.

```
SOLUTION OF SIMULTANEOUS LINEAR EQUATIONS BY GAUSSIAN ELIMINATION

    1.00000 -100.00000    0.00000    0.00000  -1.00000

    1.00000  -25.00000   -1.00000   25.00000   0.00000

    0.00000    0.00000    1.00000   -6.25000   0.00000

    0.65000   33.75000   -1.05000  -48.75000   0.00000

THE SOLUTION VECTOR IS AS FOLLOWS

    0.12500    0.01125    0.05208    0.00833

DETERMINANT =   0.6000E+04        RATIO =   0.1101E+00
```
Fig. 5.8   Gaussian elimination results for thick-walled cylinder problem

5.4.3 Solution by the Gauss–Seidel method   Before attempting to apply the Gauss–Seidel iterative process, equations (5.37) should be rearranged to try to satisfy the diagonal dominance condition given in equation (5.26). A better arrangement is

$$\begin{bmatrix} 1 & -25 & -1 & 25 \\ 1 & -100 & 0 & 0 \\ 0 & 0 & 1 & -6.25 \\ 0.65 & 33.75 & -1.05 & -48.75 \end{bmatrix} \begin{bmatrix} Y_1 \\ Z_1 \\ Y_2 \\ Z_2 \end{bmatrix} = \begin{bmatrix} 0 \\ -1 \\ 0 \\ 0 \end{bmatrix} \qquad (5.38)$$

although only the second and fourth of these equations satisfy the condition.

```
C  SOLUTION OF SIMULTANEOUS LINEAR EQUATIONS BY GAUSS-SEIDEL ITERATION.
C
       DIMENSION  A(20,20),B(20),X(20)
    1  READ(5,51) NEQN,ORELAX,TOLER,IFREQ,NCYCLE
   51  FORMAT(I5,2F10.0,2I5)
C
C  TEST VALIDITY OF THE NUMBER OF EQUATIONS.
       IF(NEQN.LT.1.OR.NEQN.GT.20) STOP
C
C  INPUT THE COEFFICIENTS.
       READ(5,52) ((A(I,J),J=1,NEQN),B(I),I=1,NEQN)
   52  FORMAT(8F10.0)
       WRITE(6,61)
   61  FORMAT(68H1SOLUTION OF SIMULTANEOUS LINEAR EQUATIONS BY GAUSS-SEID
      1EL ITERATION)
       DO 2 I=1,NEQN
    2  WRITE(6,62) (A(I,J),J=1,NEQN),B(I)
   62  FORMAT(1H0,12F10.5)
       WRITE(6,63) ORELAX
   63  FORMAT(25HOOVER-RELAXATION FACTOR =,F6.3)
C
C  SET INITIAL VALUES OF THE UNKNOWNS AS ZERO.
       DO 3 I=1,NEQN
    3  X(I)=0.
C
C  APPLY THE GAUSS-SEIDEL METHOD.
       CALL  GSITER(A,B,X,20,NEQN,ORELAX,TOLER,IFREQ,NCYCLE)
C
C  OUTPUT THE RESULTS.
       WRITE(6,64)
   64  FORMAT(34H0THE SOLUTION VECTOR IS AS FOLLOWS)
       WRITE(6,62) (X(I),I=1,NEQN)
       GO TO 1
       END
```

Fig. 5.9   Main program for solving simultaneous linear equations by the Gauss–Seidel
method

Figures 5.9 and 5.10 show a main program and Gauss–Seidel subprogram
named GSITER, for the iterative solution of simultaneous linear equations. A
flow chart for GSITER is shown in Fig. 5.11.

In the main program, the DIMENSION statement allows for up to twenty
simultaneous equations to be solved. The arrays A, B and X are used to store the
elements of matrices A, B and X defined in equation (5.3). The first executable
statement causes the number of equations to be read into NEQN, the over-
relaxation factor into ORELAX, convergence tolerance into TOLER, output
frequency into IFREQ, and maximum number of cycles of iteration into
NCYCLE. If the value of IFREQ is ten, for example, data indicating the progress
of the iteration will be written out after every ten cycles. A test is applied to
NEQN both to prevent array dimensions being exceeded, and to provide a means
of terminating execution with a blank card. Then the elements of A and B are
read in, and immediately written out following a program title. The value of the
over-relaxation factor is also written out. Of the other input data, only TOLER
and NCYCLE are written out explicitly, by the subprogram.

The values of the unknowns are set to zero before the subprogram GSITER is
called. When control of execution returns to the main program, the final
(converged) solutions are written out.

5.4.4 The Gauss–Seidel subprogram The arguments of the subprogram
GSITER shown in Fig. 5.10 include the coefficient and solution arrays A, B and
X, together with the variable NDIM which is used to dimension these arrays. The
argument MEQN enters the number of equations to be solved, and its value is
assigned to the local variable NEQN. The remaining four arguments enter the
iteration control parameters.

```
      SUBROUTINE  GSITER(A,B,X,NDIM,MEQN,ORELAX,TOLER,IFREQ,NCYCLE)
C
C  SUBROUTINE FOR SOLVING SIMULTANEOUS LINEAR EQUATIONS BY THE
C  GAUSS-SEIDEL ITERATIVE METHOD.
C
      DIMENSION  A(NDIM,NDIM),B(NDIM),X(NDIM)
      NEQN=MEQN
      IF(NEQN.LE.NDIM) GO TO 1
      WRITE(6,61)
   61 FORMAT(33HOSTOP - DIMENSION ERROR IN GSITER)
      STOP
C
C  SET UP ITERATION LOOP.
    1 IF(IFREQ.NE.0) WRITE(6,62)
   62 FORMAT(3X,4HITER,3X,5HERROR,8X,15HSOLUTION VECTOR)
      DO 5 ITER=1,NCYCLE
      SUMX=0.
      SUMDX=0.
C
C  OBTAIN NEW ESTIMATE FOR EACH UNKNOWN IN TURN.
      DO 3 I=1,NEQN
      DELTAX=B(I)
      DO 2 J=1,NEQN
    2 DELTAX=DELTAX-A(I,J)*X(J)
      DELTAX=DELTAX/A(I,I)
      SUMDX=SUMDX+ABS(DELTAX)
      X(I)=X(I)+DELTAX*ORELAX
    3 SUMX=SUMX+ABS(X(I))
C
C  DEFINE THE RELATIVE ERROR.
      ERROR=SUMDX/SUMX
C
C  TEST FOR SEVERE DIVERGENCE.
      IF(ERROR.LT.1..OR.ITER.LE.5) GO TO 4
      WRITE(6,63) ITER
   63 FORMAT(27HODIVERGENCE IN GSITER AFTER,I5,7H CYCLES)
      RETURN
C
C  TEST FOR CONVERGENCE.
    4 IF(ERROR.LT.TOLER) GO TO 6
C
C  OUTPUT PROGRESS INFORMATION EVERY IFREQ CYCLES, UNLESS IFREQ=0.
      IF(IFREQ.EQ.0) GO TO 5
      IF(MOD(ITER,IFREQ).EQ.0) WRITE(6,64) ITER,ERROR,(X(I),I=1,NEQN)
   64 FORMAT(1X,I5,E12.4,10F10.5/18X,10F10.5)
    5 CONTINUE
C
C  NORMAL EXIT FROM ITERATION LOOP INDICATES FAILURE TO CONVERGE.
      WRITE(6,65) NCYCLE
   65 FORMAT(31HONO CONVERGENCE IN GSITER AFTER,I5,7H CYCLES)
      RETURN
C
C  OUTPUT NUMBER OF ITERATIONS AND TOLERANCE FOR CONVERGED ITERATION.
    6 WRITE(6,66) TOLER,ITER
   66 FORMAT(38HOITERATION CONVERGED TO A TOLERANCE OF,E12.4,
    1          6H AFTER,I5,7H CYCLES)
      RETURN
      END
```

Fig. 5.10   Subprogram for the Gauss–Seidel method

The first action of the subprogram is to test the acceptability of the value of NEQN. If IFREQ is nonzero (f is the equivalent parameter in Fig. 5.11), a heading is written out in preparation for the data that are to be output during the iteration process. A zero value of IFREQ implies that no such results are required.

The iteration loop is set up with ITER as a counter: m is the equivalent parameter in Fig. 5.11 and Section 5.2.1. The argument NCYCLE provides the upper limit for ITER, and is shown as $\hat{m}$ in Fig. 5.11. The variables SUMX and

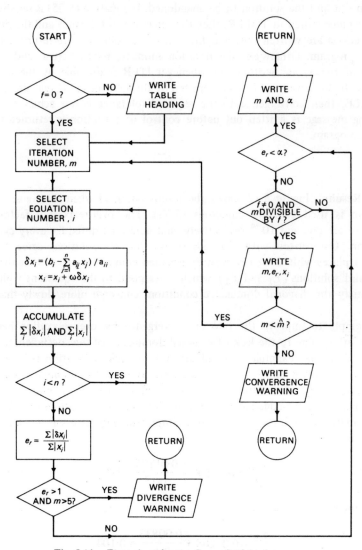

Fig. 5.11   Flow chart for the Gauss–Seidel subprogram

SUMDX, which are used to accumulate the summed magnitudes of x_i and δx_i required for the convergence test defined in equation (5.25), are set to zero at the beginning of each iteration. For each equation in turn, the variable DELTAX is used to accumulate the value of δx_i as defined by equations (5.28), from which the new value of x_i is obtained with the aid of the over-relaxation factor, ORELAX.

At the end of each iteration cycle, the relative error is calculated and stored in the variable ERROR. Severe divergence is detected if this error is greater than one after the first five cycles of iteration, and causes a warning message to be written out and the solution to be abandoned. If equation (5.25) is satisfied for the tolerance stored in TOLER, then the value of this tolerance and the number of iterations are written out, and the converged solutions are returned to the calling program. Otherwise, the iteration number, relative error, and current values of the x_i are written out whenever ITER is divisible by the nonzero IFREQ. If the iteration loop is completed by the value of ITER reaching that of NCYCLE, then convergence to the required tolerance is not achieved, and a warning message is written out before control of execution is returned to the calling program.

5.4.5 Results   Figure 5.12 shows the results obtained when the Gauss–Seidel program is used to solve equations (5.38). The over-relaxation factor and tolerance are one and 10^{-4} respectively, and data are printed for every cycle of iteration. Comparison with Fig. 5.8 shows that the solutions agree to five decimal places with those obtained by gaussian elimination. Note that the first and third solutions (Y_1 and Y_2), which correspond to the equations which do not satisfy the diagonal dominance condition, converge more slowly than the others.

Attempts to improve the rate of convergence by over-relaxation have an adverse effect, due to the lack of diagonal dominance of the coefficient matrix. In this particular example, a small amount of under-relaxation is beneficial. Using the optimum value of ω of approximately 0·90, the number of iteration cycles required is reduced from twenty to eleven, for the same tolerance.

Having found the constants in the stress distributions, equations (5.34), it is a simple matter to find the stresses. For example, the hoop stress at the innermost surface is

$$\frac{\sigma_{\theta\theta}}{P} = 0·12500 + \frac{0·01125}{0·1^2} = 1·25$$

and at the outermost surface is

$$\frac{\sigma_{\theta\theta}}{P} = 0·05208 + \frac{0·00833}{0·4^2} = 0·104$$

SOLUTION OF SIMULTANEOUS LINEAR EQUATIONS BY GAUSS–SEIDEL ITERATION

```
1.00000 -25.00000  -1.00000   25.00000   0.00000

1.00000-100.00000   0.00000    0.00000  -1.00000

0.00000    0.00000   1.00000   -6.25000   0.00000

0.65000   33.75000  -1.05000  -48.75000   0.00000
```

OVER–RELAXATION FACTOR = 1.000

ITER	ERROR	SOLUTION VECTOR			
1	0.1000E+01	0.00000	0.01000	0.00000	0.00692
2	0.8778E+00	0.07692	0.01077	0.04327	0.00755
3	0.2733E+00	0.12377	0.01124	0.04718	0.00841
4	0.6162E-01	0.11778	0.01118	0.05259	0.00818
5	0.5879E-01	0.12763	0.01128	0.05110	0.00841
6	0.3302E-01	0.12282	0.01123	0.05255	0.00828
7	0.2219E-01	0.12628	0.01126	0.05174	0.00837
8	0.1401E-01	0.12415	0.01124	0.05229	0.00831
9	0.9014E-02	0.12554	0.01126	0.05195	0.00835
10	0.5778E-02	0.12465	0.01125	0.05217	0.00832
11	0.3701E-02	0.12522	0.01125	0.05203	0.00834
12	0.2374E-02	0.12486	0.01125	0.05212	0.00833
13	0.1521E-02	0.12509	0.01125	0.05206	0.00834
14	0.9755E-03	0.12494	0.01125	0.05210	0.00833
15	0.6251E-03	0.12504	0.01125	0.05207	0.00833
16	0.4008E-03	0.12498	0.01125	0.05209	0.00833
17	0.2569E-03	0.12502	0.01125	0.05208	0.00833
18	0.1647E-03	0.12499	0.01125	0.05209	0.00833
19	0.1056E-03	0.12501	0.01125	0.05208	0.00833

ITERATION CONVERGED TO A TOLERANCE OF 1.0000E-04 AFTER   20 CYCLES

THE SOLUTION VECTOR IS AS FOLLOWS

```
0.12500    0.01125    0.05208    0.00833
```

Fig. 5.12   Gauss–Seidel results for thick-walled cylinder problem

In practice, the purpose of using compound cylinders to contain high pressures is to reduce the range of stresses that occur in a single cylinder, thereby making better use of the material in the outer cylinder. To do this most effectively, the cylinders are made with an interference fit. This condition can be accommodated in the present method, by a suitable change in equation (5.35) for the compatability of hoop strains at the interface.

5.5   Comparison of Direct and Iterative Methods

Two main methods for solving simultaneous linear equations have been presented: direct gaussian elimination and the iterative Gauss–Seidel technique. While two practical examples have been studied, the most important engineering applications are not considered until Chapters 6 and 7. Nevertheless, it is worth reviewing the advantages and disadvantages of the two methods.

Ill-conditioned equations are difficult to solve by any method, and are considered in Section 5.6. Gaussian elimination is applicable to a much wider range of problems than the Gauss–Seidel method, because of the latter's requirement of diagonal dominance. There are, however, many important practical examples of equations which satisfy this condition. If both methods are applicable, they may be compared in terms of computing time and storage requirements. As discussed in Section 2.2, these are very important economic considerations, particularly when very large sets of equations are to be solved.

5.5.1 Computing time   The most effective way to compare computing time is to count the number of arithmetic operations involved in the parts of the algorithms that are used repeatedly. It is necessary to distinguish between divisions, multiplications and additions (or subtractions) because the ratio between the times for these operations depend on the particular computer. For the gaussian elimination method, the numbers of divisions, multiplications and additions are respectively quadratic, cubic and cubic functions of n, the number of equations. If n is large, the computing time is dominated by $\frac{1}{3}n^3$ multiplications, and the same number of additions.

If Cramer's rule were to be used, then, as shown in Section 5.1.1, $n + 1$ determinants would have to be evaluated. Since each such evaluation involves of the order of $\frac{1}{3} n^3$ operations if the elimination method is used, the total number of operations is of the order of $\frac{1}{3} n^4$. Other methods of evaluating determinants, such as expansion using cofactors, are even less efficient, and Cramer's rule is never used for large sets of equations.

In the Gauss–Seidel process, the numbers of arithmetic operations are of the order of n divisions, n^2 multiplications and n^2 additions per cycle of iteration. If m iterations are required for convergence, the computing time is dominated by mn^2 multiplications, and the same number of additions. Consequently, the Gauss–Seidel method is only competitive with gaussian elimination in terms of computing time if $m \leqslant \frac{1}{3}n$. This condition is rarely obtained in practice.

So far it has been assumed that the coefficient matrix is 'dense', that is, with very few zero elements. In many practical examples involving large sets of equations, however, the coefficient matrices are sparse, which allows considerable savings of computing time to be made, particularly for the Gauss–Seidel process. Considering equation (5.28), if there are only p nonzero elements a_{ij} per row of the matrix, then only about pn multiplications and the same number of additions are required per cycle of iteration. Thus, the iterative method is more efficient than the present elimination technique if

$$mp < \frac{1}{3} n^2 \qquad (5.39)$$

This condition is studied in Section 6.6, which presents an example of the solution of partial differential equations by a finite difference method. In that case p is only five, while n may be of the order of hundreds. Equation (5.39) is derived on the assumption that no reductions in the amount of arithmetic involved in the elimination method can be achieved with sparse matrices. While such savings are sometimes possible, they are generally difficult to program. An important exception to this rule is the case of equations with a tridiagonal coefficient matrix, discussed in Section 5.1.4.

5.5.2 Storage requirements   When the elimination process is applied to sparse matrices they tend to fill up rapidly with nonzero elements. Apart from the failure to reduce the amount of arithmetic involved this also means that few savings in the amount of core store used can be achieved. A core store of n^2 words is required for the coefficient matrix when the normal gaussian elimination method is used, whereas only *pn* words are needed for the Gauss–Seidel method if the matrix is sparse. Even fewer words are required for the iterative method when the coefficients follow a regular pattern, as in the case of finite difference equations demonstrated in Section 6.6.

As indicated in Section 2.2, the cost of computing is based on both the time involved and the size of fast core store required. The Gauss–Seidel method is often preferred on both counts for the purpose of solving large sets of sparse equations. Indeed the gaussian elimination method in its present form may be impossible to use because of the limited size of the fast store. If five hundred equations are to be solved, for example, a quarter of a million words are required to store the coefficient matrix alone. There are, however, a number of techniques, such as matrix partitioning and the use of backing stores (see Section 1.1), for reducing the amount of storage required, but these are beyond the scope of this book.

5.5.3 Roundoff errors   An important advantage of iterative methods is that the roundoff errors in the final solutions are merely those incurred in the last cycle of iteration, and are therefore negligible. This is because the approximate values used in the last iteration may be regarded as initial guesses, irrespective of the process used to obtain them.

The importance of roundoff errors in the gaussian elimination method has already been mentioned in Section 5.1.2, in connection with the need to apply partial pivoting. If this is done, then roundoff errors are only likely to be significant if the equations are ill conditioned, as described in Section 5.6. Since roundoff errors are cumulative in the elimination process, they become more important as the number of equations is increased.

5.5.4 Conclusions   If a particular set of simultaneous linear equations is not diagonally dominant, then the convergence of the Gauss–Seidel method is uncertain, and gaussian elimination should normally be used. In practical examples, such as those presented in this chapter, comparatively small sets of relatively dense equations are likely to lack diagonal dominance, and are best solved by elimination. As systems of equations increase in size, they generally become increasingly sparse. In many practical examples they are also diagonally dominant. Hence, provided a reasonable rate of convergence can be achieved, the Gauss–Seidel method is to be preferred on the grounds of minimum core storage, roundoff errors, and probably computing time as well.

5.6 Ill-conditioned Equations

In Section 5.1.5, the term ill-conditioned was applied to sets of equations where the determinant of the coefficient matrix, $|A|$, is small compared to the coefficients. This may cause large errors in the solutions, because such a value of $|A|$ results from taking small differences between relatively large numbers. Both roundoff errors and inherent errors in the coefficients assume a much greater significance.

The extreme case of ill conditioning is when the determinant is zero, and the set of equations is singular. An example of singularity was presented in Section 5.3, where the equations were not all independent. In numerical computations it is difficult to distinguish between singular and extremely ill-conditioned equations, because of roundoff errors in the arithmetic.

Most practical examples of ill-conditioned equations are unsuitable for solution by the Gauss–Seidel method. The use of partial pivoting in the gaussian elimination procedure is essential if acceptable accuracy is to be maintained. The tests for ill conditioning used in the elimination program, Figs. 5.2 and 5.3, have already been outlined. A mean coefficient magnitude, $\bar{a}$, is evaluated according to equation (5.30). Since $|A|$ involves the product of n coefficients, it is appropriate to compare its magnitude with that of $\bar{a}^n$, by means of the ratio R defined in equation (5.32). The set of equations can be said to be ill-conditioned if R is small compared to one. Relative changes in one or more of the coefficients of the order of R can have large effects on $|A|$, and hence on the solutions. Consequently, R should always be several orders of magnitude greater than the likely maximum roundoff error. For example, when using a computer with twelve decimal digit precision, solutions obtained when R is of the order of 10^{-8} or less should be treated with extreme caution.

In the gaussian elimination program, tests for singularity or severe ill-conditioning are applied to avoid execution errors. A zero value of $|A|$ can only be obtained if one or more of the pivotal elements is zero, according to equation (5.15). Partial pivoting normally avoids zero pivotal values, but if the set of equations is singular, at least one zero will still be found. The effect of partial pivoting is to delay finding the zeros until the end of the elimination process, and it is usually $a_{nn}^{(n)}$ which is zero, or in practice very small, due to roundoff errors. This occurred in the example in Section 5.3, where effectively, two identical equations were used. To take an extreme example, however, if an attempt to solve n identical equations is made, then all pivotal elements after the first would be zero. In the program, all pivotal elements are tested, not as being exactly zero, but as having magnitudes less than 10^{-8} times the mean coefficient.

While roundoff errors do limit the degree of accuracy obtainable in the solution of ill-conditioned linear equations, there are methods for improving the solutions. Such methods are discussed by, for example, Conte (1965), Ralston (1965) or Williams (1972), and are particularly important when using a computer with a comparatively short word length. The use of double precision arithmetic, mentioned in Section 1.2.2, can also be beneficial.

5.7   Case Study: Currents in an Electrical Circuit Near Resonance

Figure 5.13 shows an electrical circuit having inductance and capacitance, but no resistance. The potential $V \sin \omega t$ alternates with respect to time t with an angular frequency ω and amplitude V. L_1 and L_2 are inductances, C_1 and C_2 are capacitances, and i_1, i_2 and i_3 are the currents in the circuit.

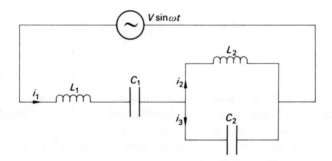

Fig. 5.13   An electrical circuit

As in the example studied in Section 5.3, equations may be written down for the conservation of currents at nodes of a network, and the summation of potential differences round closed loops. Hence, the relationship between the currents is

$$i_1 - i_2 - i_3 = 0 \qquad (5.40)$$

and, for the route followed by i_1 and i_2

$$L_1 \frac{di_1}{dt} + \frac{q_1}{C_1} + L_2 \frac{di_2}{dt} = V \sin \omega t \qquad (5.41)$$

where q_1 is the electrical charge on the capacitance C_1. Since $i_1 = dq_1/dt$, it is convenient to differentiate this equation to eliminate q_1 as follows

$$L_1 \frac{d^2 i_1}{dt^2} + \frac{i_1}{C_1} + L_2 \frac{d^2 i_2}{dt^2} = \omega V \cos \omega t \qquad (5.42)$$

Similarly, for the route followed by i_1 and i_3

$$L_1 \frac{d^2 i_1}{dt^2} + \frac{i_1}{C_1} + \frac{i_3}{C_2} = \omega V \cos \omega t \qquad (5.43)$$

In the steady state, the three currents alternate with the same frequency as the applied potential. Since in circuits containing only inductance and capacitance the currents are $90°$ out of phase with the applied potential, they may be represented by $i_1 = I_1 \cos \omega t$, $i_2 = I_2 \cos \omega t$ and $i_3 = I_3 \cos \omega t$. Substituting these expressions into equations (5.40), (5.42) and (5.43), the following set of linear equations is obtained for the amplitudes of the currents

$$\begin{bmatrix} \dfrac{1}{\omega C_1} - \omega L_1 & -\omega L_2 & 0 \\[2ex] \dfrac{1}{\omega C_1} - \omega L_1 & 0 & \dfrac{1}{\omega C_2} \\[2ex] 1 & -1 & -1 \end{bmatrix} \begin{bmatrix} I_1 \\[2ex] I_2 \\[2ex] I_3 \end{bmatrix} = \begin{bmatrix} V \\[2ex] V \\[2ex] 0 \end{bmatrix} \qquad (5.44)$$

A condition known as resonance occurs in an alternating current circuit when its impedance approaches zero. The currents tend to infinity, although in real circuits the presence of even a small amount of resistance restricts them to finite, but large, values. Resonance occurs in the present circuit when the coefficient matrix in equations (5.44) is singular.

5.7.1 Problem specification   In the circuit shown in Fig. 5.13, the amplitude of the applied potential is 300 V, the inductances are $L_1 = 0.60$ H and $L_2 = 0.20$ H, while the capacitances are $C_1 = 1$ μF and $C_2 = 2.5$ μF. The effect on the current amplitudes is to be examined as the angular frequency approaches 1000 rad/s, one of the resonant values. Also, the effect of a small inherent error in the specified value of L_1 on the currents near resonance is to be determined.

5.7.2 Solution   Figure 5.14 shows a main program for solving this problem with the aid of the gaussian elimination subprogram ELIMIN, shown in Fig. 5.3. The variables V, L1, L2, C1 and C2 are used to store the values of voltage,

```
C   PROGRAM TO FIND ELECTRICAL CURRENTS NEAR RESONANCE.
C
        DIMENSION  A(3,4)
        REAL  I(3),L1,L2
        WRITE(6,61)
 61     FORMAT(35H1ELECTRICAL CURRENTS NEAR RESONANCE)
C
C   INPUT THE CIRCUIT PARAMETERS.
  1     READ(5,51) V,L1,L2,C1,C2
 51     FORMAT(5F10.0)
        IF(V.EQ.0.) STOP
        WRITE(6,62) V,L1,L2,C1,C2
 62     FORMAT(4HOV =,F6.1,7H,   L1 =,F5.2,7H,   L2 =,F5.2,7H,   C1 =,E11.3,
     1  7H,  C2 =,E11.3 // 71H      OMEGA       DET(A)       RATIO        I1
     2            I2              I3       )
C
C   INPUT THE ANGULAR FREQUENCY.
  2     READ(5,51) OMEGA
        IF(OMEGA.EQ.0.) GO TO 1
C
C   DEFINE THE COEFFICIENTS IN THE LINEAR EQUATIONS.
        A(1,1)=1./(OMEGA*C1)-OMEGA*L1
        A(1,2)=-OMEGA*L2
        A(1,3)=0.
        A(1,4)=V
        A(2,1)=A(1,1)
        A(2,2)=0.
        A(2,3)=1./(OMEGA*C2)
        A(2,4)=V
        A(3,1)=1.
        A(3,2)=-1.
        A(3,3)=-1.
        A(3,4)=0.
C
C   SOLVE THE EQUATIONS BY GAUSSIAN ELIMINATION.
        CALL  ELIMIN(A,I,3,3,4,DETA,RATIO)
        IF(DETA.EQ.0.) GO TO 3
C
C   OUTPUT THE RESULTS.
        WRITE(6,63) OMEGA,DETA,RATIO,I
 63     FORMAT(1X,F10.5,5E12.4)
        GO TO 2
  3     WRITE(6,64) OMEGA
 64     FORMAT(1X,F10.5,33H VERY ILL-CONDITIONED OR SINGULAR)
        GO TO 2
        END
```

Fig. 5.14   Main program for finding electrical currents near resonance

inductances and capacitances defined in the physical problem. The array A is used to store the coefficients displayed in equations (5.44), while the array I is used for the three current amplitudes (note that a subscripted variable can be dimensioned in a type statement).

A program title is first written out, after which the voltage and other circuit parameters are read in. These data are immediately written out with suitable labels, followed by a table heading for the subsequent results. The angular frequency is then read into the variable OMEGA and is used to define the elements of A. Note that because ELIMIN destroys these elements during the elimination process, it is necessary to redefine even the zero values each time a new set of equations is assembled. The subprogram is then called with the relevant arguments.

When control of execution returns to the main program, the first action is to test whether the value of $|A|$, stored in the variable DETA, is exactly zero. As explained previously, this condition is deliberately imposed in ELIMIN if a zero or nearly zero pivotal element is found during elimination, which occurs when the set of equations is singular or very ill conditioned. Consequently, a warning message is written out if DETA is zero. Otherwise, the frequency and current amplitudes are written out, together with the value of $|A|$ and the ratio R defined in equation (5.32). Further values of frequency are then read in until a blank card is encountered, whereupon a new set of circuit parameters is read. Execution is terminated by two successive blank data cards.

5.7.3 Results Figure 5.15 shows some results obtained from the program shown in Fig. 5.14. Two tables are printed, the first one being for the circuit parameters defined in the problem specification. A range of angular frequencies is considered, of the form $1000 + \epsilon$ rad/s with $\epsilon = 10^2, 10^1, 10^0, \ldots, 10^{-5}$ rad/s, and finally zero. The effect is therefore to approach the resonant frequency. Despite the very small changes in frequency, say when ω is less than 1001 rad/s, there are very large changes in the amplitudes of the three currents, due to the ill conditioning of the equations. The values of both $|A|$ and R decrease rapidly with ϵ. Even at a frequency of 1100 rad/s, the equations are moderately ill conditioned in the sense that R is substantially less than one. When the magnitude of this ratio reaches the order of 10^{-6}, the equations are becoming dangerously ill conditioned from the point of view of roundoff errors. For the last three frequencies, the elimination procedure is stopped by the

```
ELECTRICAL CURRENTS NEAR RESONANCE

V = 300.0,   L1 = 0.60,   L2 = 0.20,   C1 =  1.000E-06,   C2 =  0.250E-05

   OMEGA        DET(A)        RATIO          I1            I2             I3
1100.00000  -0.4422E+05  -0.2525E-01  -0.9744E+00  -0.2467E+01   0.1492E+01
1010.00000  -0.5470E+04  -0.1553E-02  -0.1064E+02  -0.2172E+02   0.1108E+02
1001.00000  -0.5587E+03  -0.1486E-03  -0.1071E+03  -0.2146E+03   0.1075E+03
1000.10000  -0.5599E+02  -0.1479E-04  -0.1071E+04  -0.2143E+04   0.1072E+04
1000.01000  -0.5600E+01  -0.1478E-05  -0.1071E+05  -0.2143E+05   0.1071E+05
1000.00100  -0.5600E+00  -0.1478E-06  -0.1071E+06  -0.2143E+06   0.1071E+06
1000.00010  VERY ILL-CONDITIONED OR SINGULAR
1000.00001  VERY ILL-CONDITIONED OR SINGULAR
1000.00000  VERY ILL-CONDITIONED OR SINGULAR

V = 300.0,   L1 = 0.61,   L2 = 0.20,   C1 =  1.000E-06,   C2 =  0.250E-05

   OMEGA        DET(A)        RATIO          I1            I2             I3
1100.00000  -0.4580E+05  -0.2781E-01  -0.9408E+00  -0.2382E+01   0.1441E+01
1010.00000  -0.7429E+04  -0.2206E-02  -0.7835E+01  -0.1599E+02   0.8157E+01
1001.00000  -0.2555E+04  -0.7093E-03  -0.2342E+02  -0.4693E+02   0.2351E+02
1000.10000  -0.2056E+04  -0.5669E-03  -0.2918E+02  -0.5837E+02   0.2919E+02
1000.01000  -0.2006E+04  -0.5528E-03  -0.2992E+02  -0.5983E+02   0.2992E+02
1000.00100  -0.2001E+04  -0.5513E-03  -0.2999E+02  -0.5998E+02   0.2999E+02
1000.00010  -0.2000E+04  -0.5512E-03  -0.3000E+02  -0.6000E+02   0.3000E+02
1000.00001  -0.2000E+04  -0.5512E-03  -0.3000E+02  -0.6000E+02   0.3000E+02
1000.00000  -0.2000E+04  -0.5512E-03  -0.3000E+02  -0.6000E+02   0.3000E+02
```

Fig. 5.15   Electrical currents near resonance

presence of an unacceptably small pivotal element. Note that while the current amplitudes increase rapidly near resonance, the ratios between them are constant, with $2I_1 = I_2 = -2I_3$.

The results printed in the second table are for the same input data, with the exception that L_1 is increased slightly to 0·61 H. The effect on the current amplitudes is dramatic, and they are reduced by many orders of magnitude for a particular frequency. For example, with $\epsilon = 10^{-3}$ rad/s the effect of the 1·7 per cent increase in L_1 is to reduce the currents by a factor of about 3600. This highlights one of the main difficulties in problems involving ill-conditioned equations. While it is possible to guard against significant roundoff errors, if there are inherent errors in the data, say in the form of tolerances on measured quantities, then it is impossible to avoid very large errors in the solutions obtained.

5.8   Eigenvalue Problems

So far in this chapter, attention has been confined to nonhomogeneous linear equations, where at least one element of the vector B in equation (5.3) is nonzero. On the other hand, homogeneous equations normally have the trivial solution $X = 0$ (the numerators in equations (5.6) are all zero). If A is singular, however, the unknowns are indeterminate, being defined in equations (5.6) by the ratio zero over zero. A singular coefficient matrix is equivalent to having at least one pair of the equations identical, so that there are insufficient independent equations to define the unknowns uniquely. Nevertheless, simple relationships between them can be obtained. For example, if in a set of n equations $n - 1$ of them are independent, then $n - 1$ of the unknowns may each be obtained as a simple multiple of the nth.

Eigenvalue problems occur when nontrivial solutions are required to sets of homogeneous equations. Some variable parameter in the equations must be adjusted to make the coefficient matrix singular, giving an eigenvalue of that parameter. The corresponding ratios between the unknowns constitute the eigenvector.

Some of the most important eigenvalue problems that occur in engineering are associated with electrical or mechanical oscillations. The case study presented in Section 5.7 provides an example: the resonant frequency is an eigenvalue. In that particular example the equations are nonhomogeneous, and infinite currents are obtained at resonance. It was noted, however, that near resonance the ratios between the currents are constant, corresponding to the eigenvector relationship. Eigenvalue problems are also encountered in the buckling of structures, although they do not necessarily involve linear equations. Section 4.6 provides an example of this type in the buckling of a cantilevered shaft.

As demonstrated in Section 5.9, many eigenvalue problems give rise to sets of linear equations which can be expressed in the form

$$AX = \lambda X \qquad (5.45)$$

where λ is a variable parameter, while A and X are as defined in equation (5.3). Alternatively

$$(A - \lambda I)X = 0 \qquad (5.46)$$

where I is the unit matrix (a square matrix with zero elements, except for ones on the diagonal). The problem is to determine the eigenvalues of λ which make the value of the determinant $|A - \lambda I|$ zero.

For a 3 x 3 coefficient matrix, this condition may be written in full as

$$\begin{vmatrix} (a_{11} - \lambda) & a_{12} & a_{13} \\ a_{21} & (a_{22} - \lambda) & a_{23} \\ a_{31} & a_{32} & (a_{33} - \lambda) \end{vmatrix} = 0 \qquad (5.47)$$

which is a cubic equation in λ, whose roots are the three eigenvalues. For an $n \times n$ matrix there are n eigenvalues, although they are not necessarily all different. Associated with each eigenvalue is an eigenvector which defines the ratios between the x_i.

In principle it is possible to solve the polynomial equation obtained by expanding the determinant of the form shown in equation (5.47). For reasons of computing time and accuracy, however, this approach is generally unsuitable for matrices larger than about 2 x 2. Other more accurate and efficient numerical methods have been developed. For present purposes, only the simplest of these are considered, and the case study presented in Section 5.9 is concerned with a 3 x 3 coefficient matrix. More general methods for larger matrices are described by, for example, Bickley and Thompson (1964), Ralston (1965) and Williams (1972).

5.8.1 Finding the largest eigenvalue Provided the eigenvalue of largest modulus is a real number, it may be determined by a simple iterative technique. If the eigenvalues, λ_m, of an $n \times n$ matrix are all different, then the corresponding eigenvectors, X_m, are also different, and any arbitrary vector (with n elements) can be expressed as

$$Y^{(0)} = \sum_{m=1}^{n} c_m X_m \qquad (5.48)$$

where the c_m are constants. This vector may be pre-multiplied by A to give a new vector

$$Y^{(1)} = A Y^{(0)} = \sum_{m=1}^{n} c_m A X_m = \sum_{m=1}^{n} c_m \lambda_m X_m \qquad (5.49)$$

and if the process is repeated k times

$$Y^{(k)} = A^k Y^{(0)} = \sum_{m=1}^{n} c_m \lambda_m^k X_m \tag{5.50}$$

Now if λ_1 is the largest eigenvalue, it is convenient to express the summation as follows

$$Y^{(k)} = \lambda_1^k \left(c_1 X_1 + \sum_{m=2}^{n} c_m \left(\frac{\lambda_m}{\lambda_1} \right)^k X_m \right) \tag{5.51}$$

Since $(\lambda_m/\lambda_1)^k$ (with $m \neq 1$) tends to zero as k becomes large, $Y^{(k)}$ tends to become a scalar multiple of X_1, and the ratio between corresponding elements in successive Y vectors tends to λ_1. The number of iterations required for convergence depends on the ratio between λ_1 and the next largest eigenvalue, also on the choice of $Y^{(0)}$. Normally all the elements of this initial vector are set to one.

5.8.2 Finding the other eigenvalues and eigenvectors There are various methods for finding the other eigenvalues. For example, as λ_1 and X_1 are known, A can be reduced to an $(n-1) \times (n-1)$ matrix, from which the next largest eigenvalue of A can be found by the same iterative technique. This process of *deflation* can be extended to find all the eigenvalues, although with large matrices it is difficult to maintain adequate accuracy. The smallest eigenvalue can be determined by applying the iterative technique to the inverse of A. From equation (5.45)

$$A^{-1}X = \lambda^{-1}X \tag{5.52}$$

and the largest eigenvalue of A^{-1} is the reciprocal of the smallest eigenvalue of A. It is also possible to determine the eigenvalue nearest to a particular number.

Confining attention to 3×3 matrices, the second and third eigenvalues can be found with the aid of certain properties of matrices. The *trace* of a (square) matrix is the sum of the elements on its diagonal, and this is equal to the sum of its eigenvalues

$$\lambda_1 + \lambda_2 + \lambda_3 = \text{tr}\,(A) = a_{11} + a_{22} + a_{33} \tag{5.53}$$

It can be also be shown that

$$\lambda_1^2 + \lambda_2^2 + \lambda_3^2 = \text{tr}\,(A^2) \tag{5.54}$$

Similar results hold for larger matrices and higher powers of the eigenvalues. As λ_1 is known, λ_2 and λ_3 can be obtained from

$$\lambda_2 + \lambda_3 = \text{tr}\,(A) - \lambda_1 = S_1$$

$$\lambda_2^2 + \lambda_3^2 = \text{tr}\,(A^2) - \lambda_1^2 = S_2$$

$$\lambda_{2,3} = \tfrac{1}{2}(S_1 \pm \sqrt{(2S_2 - S_1^2)}) \tag{5.55}$$

and may in general be complex numbers.

The most straightforward way of finding the eigenvectors corresponding to λ_2 and λ_3 is to solve the equations

$$(A - \lambda_2 I)X_2 = 0, \qquad (A - \lambda_3 I)X_3 = 0 \qquad (5.56)$$

Taking each set in turn, it is possible to set, say, the third element, x_3, of the eigenvector to one, and then to solve for x_1 and x_2. This gives the required ratios between the unknowns, except in the special case of $x_3 = 0$, when one of the other elements should be set to one.

5.9 Case Study: Natural Frequencies of a Mechanical System with Three Degrees of Freedom

Figure 5.16 shows a mechanical system consisting of three rigid bodies with masses m_1, m_2 and m_3, linked by three springs with stiffnesses k_1, k_2 and k_3, the end of the first spring being fixed. The bodies can move without frictional resistance on a horizontal surface. The system has three degrees of freedom in

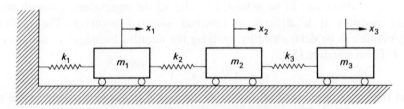

Fig. 5.16   A mechanical system with three degrees of freedom

that its configuration at any instant of time can be defined by the horizontal displacements x_1, x_2 and x_3 of the three bodies. Simplified models of this type are frequently used to represent physical problems. The equations of motion of the three bodies are as follows

$$m_1 \ddot{x}_1 = k_2 (x_2 - x_1) - k_1 x_1$$

$$m_2 \ddot{x}_2 = k_3 (x_3 - x_2) - k_2 (x_2 - x_1) \qquad (5.57)$$

$$m_3 \ddot{x}_3 = -k_3 (x_3 - x_2)$$

If the bodies are disturbed from their static equilibrium positions, they will oscillate freely with the same natural frequency ω. The accelerations are related to the corresponding displacements by $\ddot{x}_i = -\omega^2 x_i$, and equations (5.57) may therefore be arranged in matrix form as

$$\omega^2 \begin{bmatrix} m_1 & 0 & 0 \\ 0 & m_2 & 0 \\ 0 & 0 & m_3 \end{bmatrix} \begin{bmatrix} x_1 \\ x_2 \\ x_3 \end{bmatrix} = \begin{bmatrix} (k_1 + k_2) & -k_2 & 0 \\ -k_2 & (k_2 + k_3) & -k_3 \\ 0 & -k_3 & k_3 \end{bmatrix} \begin{bmatrix} x_1 \\ x_2 \\ x_3 \end{bmatrix} \quad (5.58)$$

These equations can be rearranged into the required form by inverting either the mass matrix or the stiffness matrix. Owing to the diagonal form of the mass matrix, however, it is simpler to invert, as follows

$$\begin{bmatrix} (k_1 + k_2)/m_1 & -k_2/m_1 & 0 \\ -k_2/m_2 & (k_2 + k_3)/m_2 & -k_3/m_2 \\ 0 & -k_3/m_3 & k_3/m_3 \end{bmatrix} \begin{bmatrix} x_1 \\ x_2 \\ x_3 \end{bmatrix} = \omega^2 \begin{bmatrix} x_1 \\ x_2 \\ x_3 \end{bmatrix} \quad (5.59)$$

These equations are similar to equations (5.45). The three eigenvalues of ω^2 provide the three natural frequencies of the system, and the corresponding eigenvectors define the modes of free vibration.

5.9.1 Problem specification

For the system shown in Fig. 5.16, $m_1 = 4m$, $m_2 = 2m$, $m_3 = m$, $k_1 = k$, $k_2 = k$ and $k_3 = \frac{1}{2}k$. The three natural frequencies are to be found in terms of k and m, together with the corresponding modes of free vibration.

5.9.2 Solution

With the specified data, equations (5.59) become

$$\begin{bmatrix} 0.5 & -0.25 & 0 \\ -0.5 & 0.75 & -0.25 \\ 0 & -0.5 & 0.5 \end{bmatrix} \begin{bmatrix} x_1 \\ x_2 \\ x_3 \end{bmatrix} = \frac{\omega^2 m}{k} \begin{bmatrix} x_1 \\ x_2 \\ x_3 \end{bmatrix} \quad (5.60)$$

From a particular eigenvalue of the coefficient matrix, λ, the natural frequency can be obtained as

$$\omega = (\lambda k/m)^{1/2} \quad (5.61)$$

Figure 5.17 shows a main program for finding the eigenvalues and eigenvectors of a 3 x 3 matrix. Figure 5.18 shows a subprogram named EIGEN for finding the largest eigenvalue and corresponding eigenvector of a (square) matrix of arbitrary size, by the iterative method described in Section 5.8.1. Figure 5.19 shows a subprogram named VECTOR for finding the eigenvector corresponding

```
C   PROGRAM TO FIND THE EIGENVALUES AND EIGENVECTORS OF A 3 X 3 MATRIX.
C
        DIMENSION  A(3,3),X(3),Y(3),B(3,3)
        REAL  LAMBDA(3)
    1   READ(5,51) ((A(I,J),J=1,3),I=1,3)
   51   FORMAT(3F10.0)
        IF(A(1,1).EQ.0..AND.A(1,2).EQ.0..AND.A(1,3).EQ.0.) STOP
        WRITE(6,61) ((A(I,J),J=1,3),I=1,3)
   61   FORMAT(47H1EIGENVALUES AND EIGENVECTORS OF A 3 X 3 MATRIX /
       1    (/1X,3F10.3))
C
C   FIND LARGEST EIGENVALUE AND CORRESPONDING EIGENVECTOR.
        CALL  EIGEN(A,X,Y,3,3,LAMBDA(1),1.E-8,50)
        WRITE(6,62)
   62   FORMAT(41HO    LAMBDA       X(1)      X(2)      X(3)  )
        WRITE(6,63) LAMBDA(1),X
   63   FORMAT(1X,4F10.3)
C
C   FIND TRACES OF MATRICES A AND A*A, HENCE SECOND AND THIRD EIGENVALUES
        TRACE1=A(1,1)+A(2,2)+A(3,3)
        DO 2 I=1,3
        DO 2 J=1,3
        B(I,J)=0.
        DO 2 K=1,3
    2   B(I,J)=B(I,J)+A(I,K)*A(K,J)
        TRACE2=B(1,1)+B(2,2)+B(3,3)
        S1=TRACE1-LAMBDA(1)
        S2=TRACE2-LAMBDA(1)**2
        ARG=2.*S2-S1**2
        IF(ARG.GT.0.) GO TO 3
        WRITE(6,64)
   64   FORMAT(20HOCOMPLEX EIGENVALUES)
        STOP
    3   ROOT=SQRT(ARG)
        LAMBDA(2)=0.5*(S1+ROOT)
        LAMBDA(3)=0.5*(S1-ROOT)
C
C   FIND SECOND AND THIRD EIGENVECTORS.
        DO 4 IL=2,3
        CALL  VECTOR(A,B,X,3,3,LAMBDA(IL))
    4   WRITE(6,63) LAMBDA(IL),X
        GO TO 1
        END
```

Fig. 5.17   Main program for finding eigenvalues and eigenvectors

to a given eigenvalue, with the aid of the gaussian elimination subprogram ELIMIN shown in Fig. 5.3.

In the main program it is necessary to dimension not only the arrays A and X for storing the coefficient matrix and solution vector, but also arrays Y and B for storing further vectors and matrices during the computation. The subscripted variable LAMBDA stores the three eigenvalues. The elements of A are first read in, and then written out with an appropriate program title. The subprogram EIGEN is called to find λ_1 and X_1, which are then written out under an appropriate table heading. The second and third eigenvalues are found as described in Section 5.8.2: the variables TRACE1 and TRACE2 are used to store the traces of A and A^2 respectively, while the array B is used for the elements of A^2. The required eigenvalues are found according to equation (5.55), with a suitable test to avoid an execution error in the event of them being complex numbers. The present program is not designed to handle complex eigenvalues. Finally, the subprogram VECTOR is called twice to find the second and third eigenvectors. Execution is terminated by a blank data card for the first row of a new coefficient matrix.

5.9.3 Subprogram for the largest eigenvalue

In the subprogram EIGEN, Fig. 5.18, the argument A enters the elements of the coefficient matrix, while X and Y are employed to store vectors used in the subprogram, the required eigenvector being returned to the calling program in array X. The argument

```
      SUBROUTINE  EIGEN(A,X,Y,NDIM,M,LAMBDA,TOLER,NCYCLE)
C
C  SUBROUTINE TO FIND THE LARGEST EIGENVALUE, AND CORRESPONDING
C  EIGENVECTOR OF THE SQUARE MATRIX A.
C
      DIMENSION  A(NDIM,NDIM),X(NDIM),Y(NDIM)
      REAL  LAMBDA
      N=M
C
C  SET ELEMENTS OF VECTOR X TO UNITY INITIALLY.
      DO 1 I=1,N
    1 X(I)=1.
C
C  SET UP ITERATION LOOP.
      WRITE(6,61)
   61 FORMAT(37HOITERATION FOR THE LARGEST EIGENVALUE /
     1 35H    ITER    LAMBDA          EIGENVECTOR)
      LAMBDA=0.
      DO 5 ITER=1,NCYCLE
C
C  FIRST OBTAIN MATRIX PRODUCT Y=A*X.
      DO 2 I=1,N
      Y(I)=0.
      DO 2 J=1,N
    2 Y(I)=Y(I)+A(I,J)*X(J)
C
C  FIND LARGEST ELEMENT OF VECTOR Y.
      YMAX=Y(1)
      DO 3 I=2,N
    3 IF(ABS(Y(I)).GT.ABS(YMAX)) YMAX=Y(I)
C
C  MAKE X THE NORMALISED FORM OF Y.
      DO 4 I=1,N
    4 X(I)=Y(I)/YMAX
C
C  FIND RELATIVE ERROR IN THE EIGENVALUE.
      ERROR=YMAX-LAMBDA
      IF(ABS(LAMBDA).GT.1.E-8) ERROR=ERROR/LAMBDA
      LAMBDA=YMAX
C
C  OUTPUT THE CURRENT VALUES OF THE EIGENVALUE AND EIGENVECTOR.
      WRITE(6,62) ITER,LAMBDA,(X(I),I=1,N)
   62 FORMAT(1X,I5,12F10.3)
C
C  TEST FOR CONVERGENCE.
      IF(ABS(ERROR).LT.TOLER) GO TO 6
    5 CONTINUE
C
C  NORMAL EXIT FROM ITERATION LOOP IMPLIES CONVERGENCE FAILURE.
      WRITE(6,63) NCYCLE
   63 FORMAT(21HONO CONVERGENCE AFTER,I3,7H CYCLES)
      STOP
    6 WRITE(6,64) ITER,TOLER
   64 FORMAT(18HOCONVERGENCE AFTER,I3,21H CYCLES, TO TOLERANCE,E12.4)
      RETURN
      END
```

Fig. 5.18   Subprogram for finding the largest eigenvalue

NDIM is used for dimensioning puproses, while M enters the actual size of the matrix, the value of which is assigned to the local variable N. The convergence tolerance and maximum number of iterations are entered via the arguments TOLER and NCYCLE respectively. In this case their values are set at 10^{-8} and 50 by the calling statement in the main program, although they could be read in from data cards.

Following the iterative procedure described in Section 5.8.1, the elements of the initial vector $Y^{(0)}$ are set to one, and are stored in array X. The iteration loop is then set up for repeatedly pre-multiplying this vector by the matrix A. At the kth iteration (ITER is the counter used in the program), for example, the elements of the new vector $Y^{(k)}$ are stored in array Y, while those of $Y^{(k-1)}$ are in X. After being normalised, the elements of Y are transferred to X in preparation for the next iteration. Normalisation is achieved by dividing each element by the largest element. Unbounded growth of the elements is thereby avoided, and the final eigenvector is obtained in its most convenient form. According to equation (5.51), the ratio between corresponding elements in successive Y vectors should converge to the required eigenvalue. Hence the value of YMAX, the largest element of array Y after each matrix multiplication, should converge to λ_1 (normalisation ensures that the corresponding element before multiplication is always one). The argument LAMBDA is used to store the eigenvalue. Iteration is stopped when the relative change in this value between successive cycles is less than the prescribed tolerance. In its present form, EIGEN writes out the values of λ_1 and X_1 after every cycle of iteration, and the total number of cycles and the tolerance when convergence is achieved.

5.9.4 Subprogram for the eigenvectors   In the subprogram VECTOR, Fig. 5.19, the argument A enters the elements of the coefficient matrix, while B is used to store the elements of $(A - \lambda I)$, and X returns the required eigenvector to the calling program. NDIM is used for dimensioning purposes, while M enters the actual size of the matrix, the value of which is assigned to the local variable N. The relevant eigenvalue is stored in the argument LAMBDA. Note that the corresponding argument in the calling statement in the main program is the appropriate single element of an array, which for convenience has the same name.

The elements of the singular matrix $(A - \lambda I)$ are first stored in array B, and the last element of the eigenvector is set to one. While the program is written for a general $n \times n$ coefficient matrix, the procedure may be demonstrated for the present 3×3 example. The effect of setting $x_3 = 1$ is

$$(A - \lambda I)X = BX = \begin{bmatrix} b_{11} & b_{12} & b_{13} \\ b_{21} & b_{22} & b_{23} \\ b_{31} & b_{32} & b_{33} \end{bmatrix} \begin{bmatrix} x_1 \\ x_2 \\ 1 \end{bmatrix} = \begin{bmatrix} 0 \\ 0 \\ 0 \end{bmatrix}$$

hence

$$\begin{bmatrix} b_{11} & b_{12} \\ b_{21} & b_{22} \end{bmatrix} \begin{bmatrix} x_1 \\ x_2 \end{bmatrix} = \begin{bmatrix} -b_{13} \\ -b_{23} \end{bmatrix} \tag{5.62}$$

Since the gaussian elimination subprogram ELIMIN is to be used to solve equations (5.62), it is only necessary to change the sign of the last column of array B, and reduce the number of equations by one. Since its last element is not necessarily the largest, the eigenvector is normalised before control of execution returns to the calling program.

```
      SUBROUTINE  VECTOR(A,B,X,NDIM,M,LAMBDA)
C
C  SUBROUTINE TO FIND THE EIGENVECTOR FOR A GIVEN EIGENVALUE.
C
      DIMENSION  A(NDIM,NDIM),B(NDIM,NDIM),X(NDIM)
      REAL  LAMBDA
      N=M
C
C  OBTAIN B=A-LAMBDA*(UNIT VECTOR).
      DO 1 I=1,N
      DO 1 J=1,N
      B(I,J)=A(I,J)
    1 IF(I.EQ.J) B(I,J)=B(I,J)-LAMBDA
C
C  SET X(N)=1, AND SOLVE FOR REMAINING ELEMENTS OF EIGENVECTOR X.
      X(N)=1.
      NM1=N-1
      DO 2 I=1,NM1
    2 B(I,N)=-B(I,N)
      CALL  ELIMIN(B,X,NM1,N,N,DET,RATIO)
C
C  NORMALISE THE EIGENVECTOR.
      XMAX=X(1)
      DO 3 I=2,N
    3 IF(ABS(X(I)).GT.ABS(XMAX)) XMAX=X(I)
      DO 4 I=1,N
    4 X(I)=X(I)/XMAX
      RETURN
      END
```

Fig. 5.19 Subprogram for finding the eigenvector corresponding to a given eigenvalue

The above method fails if the last element of the eigenvector happens to be zero: ELIMIN would detect singular equations. The program makes no allowance for this, but could do so by testing for a zero value of DET, and repeating the process with a different element set to one.

5.9.5 Results Figure 5.20 shows the results obtained for the present mechanical vibration problem. Convergence for the largest eigenvalue in eleven cycles is rapid, due to the ratio λ_2/λ_1 being substantially less than one. It should be noted, however, that the elements of the eigenvector converge very much more slowly than the eigenvalue. Even with a convergence tolerance as small as 10^{-8}, they are correct to no more than three significant figures.

EIGENVALUES AND EIGENVECTORS OF A 3 X 3 MATRIX

```
        0.500      -0.250      0.000

       -0.500       0.750     -0.250

        0.000      -0.500      0.500
```

ITERATION FOR THE LARGEST EIGENVALUE

ITER	LAMBDA	EIGENVECTOR		
1	0.250	1.000	0.000	0.000
2	0.500	1.000	-1.000	0.000
3	-1.250	-0.600	1.000	-0.400
4	1.150	-0.478	1.000	-0.609
5	1.141	-0.429	1.000	-0.705
6	1.140	-0.407	1.000	-0.747
7	1.140	-0.398	1.000	-0.766
8	1.140	-0.394	1.000	-0.774
9	1.140	-0.392	1.000	-0.778
10	1.140	-0.391	1.000	-0.780
11	1.140	-0.391	1.000	-0.780

CONVERGENCE AFTER 11 CYCLES, TO TOLERANCE   0.1000E-07

LAMBDA	X(1)	X(2)	X(3)
1.140	-0.391	1.000	-0.780
0.500	-0.500	-0.000	1.000
0.110	0.500	0.781	1.000

Fig. 5.20   Eigenvalues and eigenvectors for mechanical vibration problem

The exact solutions for the present problem are $\lambda_1 = (5 + \sqrt{17})/8 = 1\cdot140$, $\lambda_2 = \frac{1}{2}$, $\lambda_3 = (5 - \sqrt{17})/8 = 0\cdot110$ and the computed results are correct to at least three decimal places. The eigenvectors can be checked by substitution into equations (5.56). Using equation (5.61), the required natural frequencies are obtained as

$$\omega_1 = 1\cdot068 \left(\frac{k}{m}\right)^{\frac{1}{2}}, \qquad \omega_2 = 0\cdot707 \left(\frac{k}{m}\right)^{\frac{1}{2}}, \qquad \omega_3 = 0\cdot332 \left(\frac{k}{m}\right)^{\frac{1}{2}}$$

The eigenvectors define the modes of free vibration corresponding to these frequencies. For example, in the second mode the mass m_2 remains stationary, while m_3 oscillates out of phase with m_1 and with twice its amplitude.

As the mass matrix was inverted in order to arrive at equations (5.59), the largest eigenvalue corresponds to the highest natural frequency. In some practical problems, only the lowest natural frequency is required. This could be obtained directly by inverting the stiffness matrix in equations (5.58) and finding the largest eigenvalue of the resulting coefficient matrix.

6 Finite Difference Methods

The mathematical models for a wide range of engineering problems are expressed in terms of partial differential equations, which involve two or more independent variables. It is usually only when these models are simplified that the equations can be reduced to ordinary differential form. Some methods for solving ordinary differential equations were considered in Chapter 4. This chapter is mainly concerned with solving partial differential equations, and attention is concentrated on those types of equations that commonly occur in continuum mechanics problems. The concept of treating solid or fluid media as though they are continuous is one that is widely used in most branches of engineering. As the behaviour of all such media is governed by the same natural laws, the resulting mathematical equations are of very similar types, and can therefore be solved by the same methods.

With the exception of a few simple cases, partial differential equations cannot be solved analytically. While suitable numerical methods can usually be found, it is only with the advent of high-speed digital computers that such methods have become practicable. Finite difference methods are one of the most commonly used types, and involve replacing the derivatives in the equations by the appropriate difference approximations. These approximations are expressed in terms of the values of the dependent variable at a finite number of points within the solution domain. An alternative, though fundamentally very similar, approach is provided by the finite element method, which is introduced in Chapter 7.

6.1 Classification of Partial Differential Equations

Attention is confined here to second-order partial differential equations involving two independent variables, the general form of which can be expressed as

$$A \frac{\partial^2 \psi}{\partial x^2} + B \frac{\partial^2 \psi}{\partial x \partial y} + C \frac{\partial^2 \psi}{\partial y^2} + D \frac{\partial \psi}{\partial x} + E \frac{\partial \psi}{\partial y} + F\psi + G = 0 \tag{6.1}$$

where ψ is the dependent variable and x and y are the independent variables. If the coefficients A to G are functions of x and y only, the equation is said to be linear. In nonlinear equations they also depend on ψ or its derivatives. As shown by, for example, Crandall (1956), the values of these coefficients determine the type of equation, and hence the method of solution. The important parameter is

$$\lambda = B^2 - 4AC \qquad\qquad (6.2)$$

and equation (6.1) is said to be elliptic, parabolic or hyperbolic according to whether λ is negative, zero or positive. While it is possible for the type of an equation to change within the solution domain if A, B or C vary, in the majority of practical problems this does not happen.

Elliptic equations normally occur in equilibrium problems, whereas the parabolic and hyperbolic types occur in propagation problems. A distinction between equilibrium and propagation problems can be made in terms of the type of conditions applied at the boundaries of the solution domain. As illustrated in Fig. 6.1, the domain for an equilibrium problem is closed, and boundary

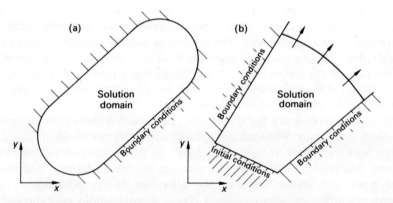

Fig. 6.1   Equilibrium and propagation problems (a) Equilibrium (boundary-value) problem with closed domain; (b) Propagation problem with open domain

conditions are prescribed around the entire boundary. Therefore, such problems are often said to be of the boundary-value type. The domain for a propagation problem, on the other hand, is open along one side. The solution is started from a side of the domain on which initial conditions are prescribed and is propagated or 'marched' towards the open side.

A similar distinction was made in Section 4.4 between initial-value (propagation) problems and boundary-value problems. In this chapter, attention is concentrated on boundary-value problems, although one example of a parabolic propagation problem is considered.

6.2 Equations Occurring in Continuum Mechanics Problems

Before considering methods for solving partial differential equations, it is appropriate to review the main types of equations occurring in a number of branches of engineering.

6.2.1 Steady laminar flow along a uniform channel
A problem concerning fluid flow along a rectangular channel was considered in Section 3.2. The velocity of flow, w, is in the z-direction normal to the cross-section shown in Fig. 3.1. The equilibrium equation for laminar flow in which only viscous and pressure forces are significant is

$$\frac{\partial}{\partial x}\left(\mu\frac{\partial w}{\partial x}\right) + \frac{\partial}{\partial y}\left(\mu\frac{\partial w}{\partial y}\right) = \frac{\partial p}{\partial z} \equiv P_z \tag{6.3}$$

where μ is viscosity and p is pressure. The pressure gradient, P_z, is independent of x and y, and the velocity is prescribed on the flow boundaries. For a newtonian fluid, equation (6.3) becomes

$$\frac{\partial^2 w}{\partial x^2} + \frac{\partial^2 w}{\partial y^2} = \nabla^2 w = \frac{P_z}{\mu} \tag{6.4}$$

where ∇^2 is the harmonic operator. Equation (6.4) is an example of Poisson's equation, whose general form is

$$\nabla^2 \psi = \phi\,(x, y) \tag{6.5}$$

Comparing this with equations (6.1) and (6.2), $A = C = 1$, $B = 0$, $\lambda = -4$: Poisson's equation is elliptic, as confirmed by the type of solution domain and boundary conditions.

6.2.2 Torsion of a prismatic bar
If a uniform elastic bar, of arbitrary cross-section in the x-y plane, is subjected to torsion, the governing equation for the stress function, χ, is

$$\nabla^2 \chi = -2G\theta \tag{6.6}$$

where G is the shear modulus, and θ is the angle of twist per unit length of the bar. Another Poisson equation is obtained, and the boundary conditions can be expressed as prescribed values of χ.

6.2.3 Deflection of a taut membrane

If a flat membrane of arbitrary shape, lying in the x-y plane, is subjected to a lateral pressure, p, the governing equation for the (small) lateral deflection, w, is

$$\nabla^2 w = p/S \tag{6.7}$$

where S is the uniform tension per unit length in the membrane.

6.2.4 Ideal fluid flow

In the ideal fluid model, only pressure and inertia forces are significant. The equilibrium equations for flow in the x-y plane reduce to the single condition

$$\nabla^2 \psi = 0 \tag{6.8}$$

where ψ is a stream function. This is an example of Laplace's equation, which is a special case of Poisson's equation. The boundary conditions are likely to involve both prescribed values of stream function, and, on some parts of the boundary, prescribed values of its first derivatives.

6.2.5 Diffusion problems

There are many physical processes which involve some form of diffusion. The fluxes in the two coordinate directions in the plane of the solution domain are

$$q_x = -k \frac{\partial \psi}{\partial x}, \qquad q_y = -k \frac{\partial \psi}{\partial y} \tag{6.9}$$

where k is the appropriate 'conductivity' property of the medium concerned, and ψ is the relevant 'potential'. Assuming that the process is steady and that there are no local gains or losses of flux, then

$$\frac{\partial q_x}{\partial x} + \frac{\partial q_y}{\partial y} = 0 \tag{6.10}$$

Provided k is constant, Laplace's equation is obtained. Some practical examples of diffusion problems are

(1) thermal conduction in a solid or fluid medium;
(2) electrical conduction in a solid or fluid medium;
(3) electrostatic potential distribution in an insulating medium;
(4) fluid flow in a porous medium, such as water in soil;
(5) neutron diffusion in a nuclear reactor.

Considering heat transfer by thermal conduction in more detail, the general equation is

$$k\nabla^2 T + g = \rho\, C_p\, \frac{\partial T}{\partial t} \tag{6.11}$$

where T is temperature, t is time, g is the rate of heat generation per unit volume, and ρ, C_p and k are the density, specific heat and thermal conductivity of the material concerned. Under steady conditions, equation (6.11) reduces to the elliptic Poisson or Laplace equation according to whether or not there is any local heat generation. Boundary conditions prescribe either T, its derivatives, or some relationship between them. The simplest example of a derivative condition is $\partial T/\partial n = 0$, where n is the direction of the outward normal to the thermally insulated boundary.

In order to reduce the unsteady case to a problem involving only two independent variables, it is convenient to assume that conduction occurs only in the x-direction, and equation (6.11) becomes

$$k \frac{\partial^2 T}{\partial x^2} - \rho C_p \frac{\partial T}{\partial t} + g = 0 \qquad (6.12)$$

Comparing this with equations (6.1) and (6.2), $A = k$, $B = 0$, $C = 0$, $\lambda = 0$: equation (6.12) is parabolic. The temperature profile in the x-direction is defined for some initial time, and the solution is propagated with time.

6.2.6 Flow in a boundary layer   A further example of a parabolic equation is provided by steady newtonian flow in a boundary layer. If the solid boundary lies in the x-z plane, and the direction of flow is parallel to the x-axis, then

$$\frac{\mu}{\rho} \frac{\partial^2 u}{\partial y^2} - \frac{1}{\rho} \frac{dp}{dx} = u \frac{\partial u}{\partial x} + v \frac{\partial u}{\partial y} \qquad (6.13)$$

where u and v are the velocity components in the x and y-directions, ρ and μ are the fluid density and viscosity respectively, and $p(x)$ is the pressure distribution in the fluid stream outside the boundary layer. The velocity v can be eliminated with the aid of the equation for mass conservation.

6.2.7 Hydrodynamic lubrication   Reynolds' equation for flow in a thin film of lubricant between moving surfaces provides an example of a more general type of elliptic equation. If V is the relative velocity, in the x-direction, between surfaces which are parallel to the x-y plane, the pressure in the lubricant is governed by

$$H^3 \nabla^2 p + \frac{\partial H^3}{\partial x} \frac{\partial p}{\partial x} + \frac{\partial H^3}{\partial y} \frac{\partial p}{\partial y} = 6 \mu V \frac{\partial H}{\partial x} \qquad (6.14)$$

where μ is the viscosity and $H(x, y)$ is the thickness of the film.

6.2.8 Biharmonic problems While attention is concentrated on second-order partial differential equations, there are a number of important practical problems that involve higher-order equations. For example, while slow viscous flow normal to the solution domain (Section 6.2.1) and two-dimensional ideal fluid flow (Section 6.2.4) are governed by Poisson and Laplace equations respectively, two-dimensional slow viscous flow is governed by

$$\nabla^4 \psi \equiv \nabla^2 (\nabla^2 \psi) = 0 \tag{6.15}$$

where ψ is a stream function. This elliptic fourth-order equation is of the so-called biharmonic type. One method of solving such an equation is to reduce it to a pair of simultaneous second-order equations

$$\nabla^2 \psi = \omega, \qquad \nabla^2 \omega = 0 \tag{6.16}$$

where ω is vorticity. Since these are of the Poisson type they can in principle be solved by the present methods. Other problems governed by biharmonic equations include plane strain and plane stress in elastic media, and the bending of elastic plates.

6.2.9 General comments The fact that so many problems involve equations of the Poisson type indicates the close analogy which exists between them, and is of considerable importance in experimental work. For example, it is possible to set up an electrical analogue of a thermal conduction problem. Similarly, predictions of channel flow velocity profiles have been obtained from deflection measurements on membranes subject to lateral pressure.

In all problems of the boundary-value type it is the boundary conditions as much as the differential equations that determine the solution. While there are many mathematical functions that satisfy the Poisson or Laplace equations, they rarely satisfy the required boundary conditions. One method of solution, however, is to employ an infinite series of such functions to satisfy the boundary conditions, as implied in Section 3.2.

In the remainder of this chapter, attention is concentrated on finite difference methods for solving the elliptic Poisson equation (6.5), and the parabolic diffusion equation (6.12).

6.3 Boundary-Value Problems with One Independent Variable

As a preliminary to solving boundary-value problems governed by the full Poisson equation, it is convenient to examine situations where the dependence on one of the variables may be neglected. If both ψ and ϕ are independent of y, equation (6.5) becomes

$$\frac{d^2 \psi}{dx^2} = \phi(x) \tag{6.17}$$

Such an ordinary differential equation can be solved by the methods outlined in Section 4.4.3. While the predictor–corrector method discussed in Section 4.4.2 could be described as a finite difference method, the term is normally reserved for methods that involve central differences.

6.3.1 Finite difference analysis Finite difference methods involve replacing derivatives by difference approximations. Continuous functions, such as ψ in equation (6.17), are thereby represented by a finite number of values at prescribed points in the solution domain. Figure 6.2 shows a uniformly

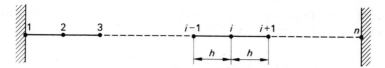

Fig. 6.2   Uniformly distributed points in a one-dimensional solution domain

distributed set of points in the one-dimensional domain. There are n points, including those on the boundaries, and the distance between successive points is h. The derivative in equation (6.17) may be expressed for a typical point i in terms of the values of ψ at adjacent points, by means of the central difference formula

$$\left(\frac{d^2\psi}{dx^2}\right)_i = \frac{\psi_{i-1} - 2\psi_i + \psi_{i+1}}{h^2} + e_T \tag{6.18}$$

This result may be obtained from Taylor's series, and the truncation error is of the order of

$$e_T \approx -\frac{h^2}{12}\left(\frac{d^4\psi}{dx^4}\right)_i \tag{6.19}$$

If this error is ignored, equation (6.18) provides an approximation to the differential equation (6.17)

$$\psi_{i-1} - 2\psi_i + \psi_{i+1} = h^2\phi_i \tag{6.20}$$

which is applicable to all internal points.

Equations for the boundary points may be obtained from the relevant boundary conditions. For example, if the value of ψ is prescribed as ψ_0 at the right-hand boundary in Fig. 6.2, then $\psi_n = \psi_0$. Derivative boundary conditions can be applied with the aid of the relevant finite difference formulae. For example, the first derivative at the left-hand boundary is given by

$$\left(\frac{d\psi}{dx}\right)_1 = \frac{\psi_2 - \psi_1}{h} + O\,(h) \tag{6.21}$$

and if this derivative is required to take the prescribed value η, then ϕ_1 is given by the approximation

$$\psi_1 = \psi_2 - \eta h \tag{6.22}$$

It should be noted, however, that the truncation error in equation (6.21) is of a different order from that in equation (6.18). A better approximation to the derivative at the boundary is

$$\left(\frac{d\psi}{dx}\right)_1 = \frac{1}{2h}\,(-3\psi_1 + 4\psi_2 - \psi_3) + O\,(h^2) \tag{6.23}$$

hence

$$\psi_1 = \frac{1}{3}\,(4\psi_2 - \psi_3 - 2\eta h) \tag{6.24}$$

Section 6.8 provides an example of a boundary condition which prescribes a relationship between ψ and its first derivative.

Using equations (6.20) and (6.22), together with the prescribed value of ψ_n, the following set of n simultaneous linear equations is obtained for the ψ_i

$$\begin{bmatrix} -1 & 1 & & & & \\ 1 & -2 & 1 & & \text{\Large 0} & \\ & 1 & -2 & 1 & & \\ & & \cdot & \cdot & \cdot & \\ \text{\Large 0} & & & 1 & -2 & 1 \\ & & & & 0 & 1 \end{bmatrix} \begin{bmatrix} \psi_1 \\ \psi_2 \\ \psi_3 \\ \cdot \\ \psi_{n-1} \\ \psi_n \end{bmatrix} = \begin{bmatrix} \eta h \\ h^2\phi_2 \\ h^2\phi_3 \\ \cdot \\ h^2\phi_{n-1} \\ \psi_0 \end{bmatrix} \tag{6.25}$$

The coefficient matrix is of the tridiagonal form indicated in equation (5.19).

6.3.2 Solution of the tridiagonal set of equations

Tridiagonal sets of linear equations are best solved by a special form of gaussian elimination, in which only the nonzero elements of the coefficient matrix need be stored. For computing purposes an $n \times 3$ array can be used, and for the general case the equations can be written in a similar notation to equation (5.2), as follows.

$$
\begin{bmatrix}
a_{12} & a_{13} & & & & 0 \\
a_{21} & a_{22} & a_{23} & & & \\
 & a_{31} & a_{32} & a_{33} & & \\
0 & & & \ddots & \ddots & \ddots \\
 & & & & a_{n1} & a_{n2}
\end{bmatrix}
\begin{bmatrix}
x_1 \\ x_2 \\ x_3 \\ \vdots \\ x_n
\end{bmatrix}
=
\begin{bmatrix}
b_1 \\ b_2 \\ b_3 \\ \vdots \\ b_n
\end{bmatrix}
\qquad (6.26)
$$

The element a_{ij} $(1 \leqslant i \leqslant n, 1 \leqslant j \leqslant 3)$ is either below, on or above the diagonal according to whether j is 1, 2 or 3. The elimination process is similar to the one described in Section 5.1.2. If the coefficients shown in equations (6.26) are given the notation $a_{ij}^{(1)}$, $b_i^{(1)}$, the modified coefficients obtained when the first equation is used to eliminate x_1 from the second equation are

$$
a_{22}^{(2)} = a_{22}^{(1)} - \theta\, a_{13}^{(1)}
$$

$$
b_2^{(2)} = b_2^{(1)} - \theta\, b_1^{(1)} \qquad (6.27)
$$

$$
\theta = a_{21}^{(1)}/a_{12}^{(1)}
$$

Similarly, after the $i-1$th elimination, the coefficients in the ith equation are

$$
a_{i2}^{(i)} = a_{i2}^{(i-1)} - \theta\, a_{i-1,3}^{(i-1)}
$$

$$
b_i^{(i)} = b_i^{(i-1)} - \theta\, b_{i-1}^{(i-1)} \qquad (6.28)
$$

$$
\theta = a_{i1}^{(i-1)}/a_{i-1,2}^{(i-1)}
$$

while $a_{i1}^{(i)}$ is reduced to zero, and $a_{i3}^{(i)}$ is unchanged. When the elimination process is complete, the unknowns can be obtained in reverse order by back substitution

$$
x_n = b_n^{(n)}/a_{n2}^{(n)} \qquad (6.29)
$$

$$
x_i = (b_i^{(i)} - a_{i3}^{(i)} x_{i+1})/a_{i2}^{(i)} \qquad (6.30)
$$

$$
i = n-1, n-2, \ldots, 1
$$

In Section 5.5.1, the number of arithmetic operations for the general gaussian elimination and Gauss–Seidel methods for solving linear equations were stated. The present method for tridiagonal sets is much more efficient as it involves approximately $2n$ divisions, $3n$ multiplications and $3n$ additions.

6.4   Case Study: Temperature Distribution in Fluid Flow

Figure 6.3 depicts the flow of a very viscous fluid between two stationary parallel flat plates, under the influence of a pressure gradient in the x-direction. The distribution of the velocity in this direction is of parabolic shape, and is given by

$$u = \frac{4U}{H^2}(yH - y^2) \qquad (6.31)$$

where U is the maximum velocity (at $y = \frac{1}{2}H$, y being the distance above the lower plate) and H is the distance between the plates. The flow is assumed to be fully developed, so that both velocities and temperatures are independent of x.

The temperature distribution in the fluid is given by the solution to equation (6.11), which for steady fully developed flow becomes

$$k\frac{d^2 T}{dy^2} = -g \qquad (6.32)$$

In this case, g is the rate of dissipation of mechanical work, which is the product of the local shear stress and shear rate, and may be obtained with the aid of equation (6.31) as

$$g = \mu\left(\frac{du}{dy}\right)^2 = \frac{16\mu U^2}{H^4}(H - 2y)^2 \qquad (6.33)$$

The lower plate is thermally insulated and the upper plate is maintained at a constant temperature T_0, so the temperature boundary conditions are

$$\frac{dT}{dy} = 0 \text{ at } y = 0, \quad T = T_0 \text{ at } y = H \qquad (6.34)$$

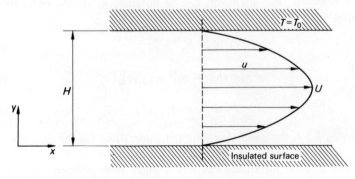

Fig. 6.3   Fluid flow between parallel plates

It is convenient to introduce the following dimensionless variables

$$Y = \frac{y}{H}, \quad T^* = \frac{k(T - T_0)}{\mu U^2} \tag{6.35}$$

and equations (6.32) and (6.34) become

$$\frac{d^2 T^*}{dY^2} = -16 (1 - 2Y)^2 \tag{6.36}$$

$$\frac{dT^*}{dY} = 0 \text{ at } Y = 0, \quad T^* = 0 \text{ at } Y = 1 \tag{6.37}$$

6.4.1 Problem specification   The dimensionless temperature profile $T^*(Y)$ is to be found by solving equations (6.36) and (6.37) by means of the finite difference procedure described in Section 6.3. The accuracy of the method is to be assessed by using from 2 to 128 steps between points in the solution domain, and comparing the computed temperature at the insulated surface, $T^*(0)$, with the analytical solution. This solution is

$$T^*(Y) = \frac{1}{3} (9 - 8Y - (1 - 2Y)^4) \tag{6.38}$$

$$T^*(0) = 2\tfrac{2}{3} \tag{6.39}$$

6.4.2 Solution   Figures 6.4 and 6.5 show respectively a main program for this problem, and a subprogram for solving the resulting tridiagonal set of linear equations. In the main program, the DIMENSION statement allows for up to 130 simultaneous equations to be solved. The arrays A and B are used to store the coefficients a_{ij} and b_i displayed in the general equations (6.26), while the array T is used for the required temperatures. After writing out an appropriate program title and table heading, the program reads in and tests the number of steps to be used in the solution domain. The number of points is one more than the number of steps, and is stored in IMAX, while the step length, h, for the dimensionless co-ordinate Y is stored in the variable H.

Equation (6.36) is analogous to equation (6.17), on which the finite difference analysis described in Section 6.3.1 was based. From equations (6.25) and (6.26), the elements of the tridiagonal coefficient matrix are

$$a_{i1} = 1, \quad a_{i2} = -2, \quad a_{i3} = 1, \quad i = 2, 3, \ldots, n-1 \tag{6.40}$$

and the corresponding elements of the vector containing the right-hand sides of the equations are

$$b_i = h^2 \phi_i = -16h^2 (1 - 2Y_i)^2 \tag{6.41}$$

```
C  PROGRAM FOR TEMPERATURE DISTRIBUTION IN FLUID FLOW (BOUNDARY-VALUE
C  PROBLEM YIELDING A TRIDIAGONAL MATRIX).
C
       DIMENSION  A(130,3),B(130),T(130)
       WRITE(6,61)
   61  FORMAT(39H1TEMPERATURE DISTRIBUTION IN FLUID FLOW /
      1 16H  NSTEP     T(0) )
    1  READ(5,51) NSTEP
   51  FORMAT(I5)
       IF(NSTEP.LE.O.OR.NSTEP.GE.130) STOP
       IMAX=NSTEP+1
       H=1./FLOAT(NSTEP)
C
C  DEFINE SECOND DERIVATIVE OF TEMPERATURE AT EACH POINT, HENCE
C  ELEMENTS OF VECTOR B, ALSO THOSE OF TRIDIAGONAL MATRIX A.
       FACT=-16.*H**2
       DO 2 I=2,NSTEP
       Y=H*FLOAT(I-1)
       B(I)=FACT*(1.-2.*Y)**2
       A(I,1)=1.
       A(I,2)=-2.
    2  A(I,3)=1.
C
C  IMPOSE FIXED BOUNDARY TEMPERATURE CONDITION.
       A(IMAX,1)=0.
       A(IMAX,2)=1.
       B(IMAX)=0.
C
C  IMPOSE BOUNDARY CONDITION AT INSULATED SURFACE.
       A(1,2)=1.
       A(1,3)=-1.
       B(1)=0.
C
C  SOLVE THE TRIDIAGONAL SET OF EQUATIONS.
       CALL  TRIDIA(A,B,T,130,IMAX)
C
C  OUTPUT THE TEMPERATURE AT THE INSULATED SURFACE.
       WRITE(6,62) NSTEP,T(1)
   62  FORMAT(1X,I5,F10.4)
       GO TO 1
       END
```

Fig. 6.4   Main program for finding the temperature distribution in fluid flow

In the program, Fig. 6.4, the common factor $-16h^2$ is stored in the variable FACT before the DO loop for defining the coefficients is entered. The required temperature boundary conditions can be imposed by defining the coefficients in the first and last equations as shown in equations (6.25), where for the present problem the prescribed boundary temperature and derivative are both zero. Having applied the boundary conditions, the main program calls the subprogram TRIDIA to solve the equations. When control of execution returns to the main program, the number of steps and the dimensionless temperature at the insulated surface are written out before a new value of NSTEP is read.

6.4.3 The tridiagonal elimination subprogram The subprogram named TRIDIA, shown in Fig. 6.5, uses the method described in Section 6.3.2 for solving a tridiagonal set of simultaneous linear equations. In the gaussian elimination subprogram described in Section 5.3.3, the vector B was treated as an extension of the coefficient matrix A. Owing to the way in which the a_{ij} have

```
C          SUBROUTINE  TRIDIA(A,B,X,NDIM,MEQN)
C
C   SUBROUTINE FOR SOLVING A TRIDIAGONAL SET OF EQUATIONS BY ELIMINATION.
C
           DIMENSION  A(NDIM,3),B(NDIM),X(NDIM)
           NEQN=MEQN
           IF(NEQN.LE.NDIM) GO TO 1
           WRITE(6,61)
   61      FORMAT(33HOSTOP - DIMENSION ERROR IN TRIDIA)
           STOP
C
C   PERFORM ELIMINATIONS.
    1      DO 2 I=2,NEQN
           FACT=A(I,1)/A(I-1,2)
           A(I,2)=A(I,2)-A(I-1,3)*FACT
    2      B(I)=B(I)-B(I-1)*FACT
C
C   USE BACK SUBSTITUTION TO SOLVE FOR THE UNKNOWNS.
           X(NEQN)=B(NEQN)/A(NEQN,2)
           DO 3 L=2,NEQN
           I=NEQN+1-L
    3      X(I)=(B(I)-A(I,3)*X(I+1))/A(I,2)
           RETURN
           END
```

Fig. 6.5   Subprogram for solving tridiagonal sets of linear equations

been renumbered, as shown in equations (6.26), there is no advantage to be gained by using the same device in the present program.

The arrays A, B, and X, which store the equation coefficients and solutions, are dimensioned with the aid of the argument NDIM. The argument MEQN enters the number of equations to be solved, and its value is assigned to the local variable NEQN. The eliminations are performed according to equations (6.28), and the solutions obtained by the back substitution process defined by equations (6.29) and (6.30).

6.4.4 Results   Figure 6.6 shows the results obtained from the program shown in Fig. 6.4 and 6.5, for the specified numbers of steps. Even when as many as 128 steps are used, the computed temperature at the insulated surface differs from the exact solution, equation (6.39), by 2·3 per cent.

The main reason for this low accuracy is that the derivative boundary condition is applied with the aid of equation (6.21), which involves a large truncation error. Equation (6.23) provides a much better approximation. At first

TEMPERATURE DISTRIBUTION IN FLUID FLOW

NSTEP	T(0)
2	0.0000
4	1.0000
8	1.7500
16	2.1875
32	2.4219
64	2.5430
128	2.6045

Fig. 6.6   Results from temperature distribution program

sight it appears, however, that this expression cannot be used in the present program, because the coefficient of ψ_3 in the first of equations (6.25) is thereby made nonzero, which violates the required tridiagonal form. This difficulty can be overcome by using the second of equations (6.25) to eliminate ψ_3 from the first to give

$$\psi_1 = \psi_2 - \tfrac{1}{2}h^2\phi_2 - \eta h \tag{6.42}$$

In the present problem, the value of the derivative is zero, and the only alteration required in the coefficients displayed in equations (6.25) is for b_1 to be changed from ηh (zero) to

$$b_1 = -\tfrac{1}{2}b_2 \tag{6.43}$$

If the relevant program statement in Fig. 6.4 is modified, the results shown in Fig. 6.7 are obtained. The agreement with the exact solution is much improved, and for 128 steps the error is reduced to 0·06 per cent.

```
TEMPERATURE DISTRIBUTION IN FLUID FLOW
    NSTEP      T(0)
      2       0.0000
      4       1.5000
      8       2.3125
     16       2.5703
     32       2.6416
     64       2.6603
    128       2.6651
```

Fig. 6.7   Temperature distribution results using improved boundary condition, equation (6.43)

As with all numerical methods for solving differential equations, the accuracy of the solution can be improved by increasing the number of points employed in the solution domain. This improvement is only attained, however, at the expense of increased computing time and storage requirements. While the present program is not a heavy user of either of these resources, in problems involving more than one independent variable, the use of 128 steps per co-ordinate direction would normally be uneconomic.

The present finite difference method is not capable of producing very accurate solutions, because its truncation error is relatively large, being of the order of h^2, as indicated by equation (6.19). If solutions are available for two or more sizes of step length, however, a form of extrapolation (sometimes called h^2 extrapolation) can be used to obtain improved estimates of the true solutions. For example, if a set of solutions $\psi_i^{(1)}$ are obtained with a step length of h, and another set $\psi_i^{(2)}$ when the step length is $\tfrac{1}{2}h$, then if ψ_i are the true solutions

$$\psi_i \approx \psi_i^{(1)} + \epsilon h^2 \approx \psi_i^{(2)} + \frac{1}{4}\epsilon h^2 \tag{6.44}$$

Table 6.1
Extrapolated values of dimensionless temperature

h^{-1}	$T(0)^{(1)}$	$T(0)^{(2)}$	$T(0)$
4	1·5000	2·3125	2·5833
8	2·3125	2·5703	2·6562
16	2·5703	2·6416	2·6654
32	2·6416	2·6603	2·6665
64	2·6603	2·6651	2·6667

where ϵ is the constant of proportionality in the error term, of the form shown in equation (6.19). Eliminating the error term

$$\psi_i \approx \frac{1}{3}\left(4\psi_i^{(2)} - \psi_i^{(1)}\right) \tag{6.45}$$

Table 6.1 shows the extrapolated values of the required dimensionless temperatures, obtained from the results shown in Fig. 6.7. The improvement in the results is considerable: using the solutions for 16 and 32 steps, plus extrapolation, the result is more accurate than using 128 steps alone. It is generally more economical in terms of both computing time and storage to obtain improved accuracy by extrapolation, rather than to increase the number of steps.

6.5   Boundary-Value Problems with Two Independent Variables

In Section 6.3, a finite difference analysis was applied to the one-dimensional form of the general Poisson equation. A similar method of solution can be used for the full equation. Further details of this and other finite difference methods are given by, for example, Crandall (1956).

6.5.1 Finite difference analysis   From equation (6.5), the cartesian form of Poisson's equation is

$$\frac{\partial^2 \psi}{\partial x^2} + \frac{\partial^2 \psi}{\partial y^2} = \phi\,(x,\,y) \tag{6.46}$$

Figure 6.8 shows part of a grid in the two-dimensional solution domain. The lines of the grid are uniformly spaced in the x and y-directions, the distances between them being h_x and h_y respectively. The points to be used in the finite difference analysis are located at the intersections of the grid lines. The counters i and j, which define the point positions in the x and y directions respectively, are convenient to use as subscripts when programming the analysis. For present

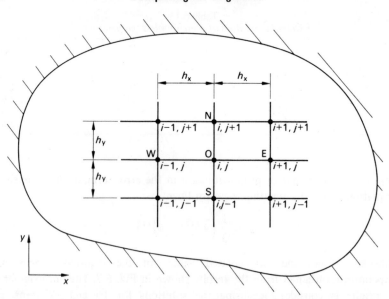

Fig. 6.8   Retangular grid in a two-dimensional solution domain

purposes, however, the local compass-point subscripts N, S, E and W for the four points adjacent to the arbitrary point subscripted O provide a more pictorial notation. Following equation (6.18), the finite difference representation of equation (6.46), for the point 'O' within the solution domain, is

$$\frac{\psi_W - 2\psi_O + \psi_E}{h_x^2} + \frac{\psi_S - 2\psi_O + \psi_N}{h_y^2} = \phi_O \qquad (6.47)$$

and the truncation error is again of the order of the square of the grid step lengths.

In virtually all finite difference methods, the grid lines are parallel to the co-ordinate axes. Therefore, using cartesian co-ordinates, the method is best suited to solving problems with rectangular solution domains, so that the rows of uniformly spaced points can terminate with points on the domain boundary. Prescribed boundary values or derivatives can then be applied as in the one-dimensional case. Irregular boundary shapes are much more difficult to accommodate, and the points near the boundary require special treatment. Such problems are often more conveniently solved by the alternative finite element method, as discussed in Section 7.4.

6.5.2 Solution of the linear equations If there are n points in the solution domain then there are n linear equations of the form of either equation (6.47), or expressing the relevant boundary conditions. As indicated in Section 5.5, such a sparse set of equations is best solved by the Gauss–Seidel method. Following equation (5.23), an improved estimate of the value of ψ at an internal point 'O' can be obtained in terms of the latest values at the surrounding points as

$$\psi_O = (\gamma_W \psi_W + \gamma_E \psi_E + \gamma_S \psi_S + \gamma_N \psi_N - \phi_O)/\gamma_O \qquad (6.48)$$

where, from equation (6.47)

$$\gamma_O = 2\left(\frac{1}{h_x^2} + \frac{1}{h_y^2}\right), \gamma_W = \gamma_E = \frac{1}{h_x^2}, \gamma_S = \gamma_N = \frac{1}{h_y^2} \qquad (6.49)$$

The over-relaxation process described in Section 5.2.3 can be applied to increase the rate of convergence. Since

$$|\gamma_O| = |\gamma_W| + |\gamma_E| + |\gamma_S| + |\gamma_N| \qquad (6.50)$$

the equality in the diagonal dominance condition expressed in equation (5.26) is satisfied for the equations determining ψ at all internal grid points. Equations defining prescribed values for ψ on the boundary satisfy the inequality in equation (5.26), while equations such as equation (6.22) for derivative boundary conditions satisfy the equality. Consequently, the coefficient matrix is diagonally dominant and the Gauss-Seidel method is convergent. It should be noted, however, that the more accurate equation (6.24) for the derivative boundary condition does not satisfy the diagonal dominance condition, but this does not normally prevent convergence.

6.6 Case Study: Fluid Flow Along a Rectangular Channel

Any of the physical problems outlined in Sections 6.2.1 to 6.2.5 could be used to illustrate the solution of Poisson's equation by definite difference method. The steady laminar flow of a viscous newtonian fluid along a uniform channel of rectangular cross-section is chosen because an accurate solution is available for comparison. This problem was outlined in Section 6.2.1, and in Section 3.2 a method of evaluating the shape factor for the volumetric flow rate, using an infinite series, was presented. This series results from the search for a mathematical function that satisfies both the Poisson equation (equation (6.4) or (3.5)), and the condition of zero velocity on the channel boundaries.

It is convenient to introduce the following dimensionless variables

$$\psi = w/\bar{w}, \quad X = x/W, \quad Y = y/W \qquad (6.51)$$

where W is the channel width as shown in Fig. 3.1, and the mean velocity, $\bar{w}$, is the ratio between the volumetric flow rate and channel area

$$\bar{w} = Q/WH \qquad (6.52)$$

Equation (6.4) can now be expressed as

$$\frac{\partial^2 \psi}{\partial X^2} + \frac{\partial^2 \psi}{\partial Y^2} = \phi = \frac{P_z W^2}{\mu \bar{w}} \tag{6.53}$$

and the volumetric flow rate is given by

$$Q = \int_0^H \int_0^W w \, dx \, dy = \bar{w} \, W^2 \int_0^{H/W} \int_0^1 \psi \, dX \, dY \tag{6.54}$$

The flow-rate shape factor is defined by equation (3.6) as

$$F = -\frac{12\mu Q}{P_z W H^3} = -\frac{12}{\phi} \left(\frac{W}{H}\right)^3 \int_0^{H/W} \int_0^1 \psi \, dX \, dY \tag{6.55}$$

If just the shape factor is required, an arbitrary constant value can be assigned to the parameter ϕ used in equations (6.53) and (6.55).

The boundary conditions for the dimensionless velocity are $\psi = 0$ on all four boundaries, as shown in Fig. 6.9a. Since the flow is symmetrical about the two dotted lines shown in this diagram, it is possible to examine only one quadrant of the solution domain. For example, Fig. 6.9b shows the shaded quadrant

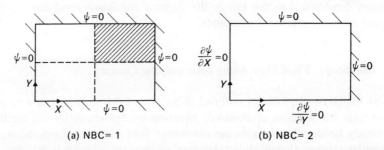

(a) NBC= 1   (b) NBC= 2

Fig. 6.9   Alternative solution domains and boundary conditions for channel flow problem

enlarged to the same size as the original domain, with boundary conditions that take account of the symmetry. If just the shape factor is required, the value obtained with the latter domain and boundary conditions should be divided by four. The potential advantage of using symmetry in this way is that greater accuracy can be obtained with the same number of points in the solution domain.

6.6.1 Problem specification

The shape factor for volumetric flow rate is to be computed by the finite difference method, for a channel depth to width ratio of

$H/W = 0.5$. Solutions are to be obtained for grids of 5 x 5, 10 x 10, and 20 x 20 steps in the solution domain, and compared with the accurate result $F = 0.6860$ found in Section 3.2. Both sets of boundary conditions shown in Fig. 6.9 are to be investigated, and optimum over-relaxation factors for the Gauss–Seidel method of solving the finite difference equations are to be found by trial and error.

6.6.2 Solution Figure 6.10 shows a main program for this problem, and Fig. 6.11 shows a subprogram named POISSN for solving Poisson's equation in a rectangular solution domain by finite difference method. In the main program, the array VELOC is used to store the dimensionless velocities and is dimensioned to allow for the use of up to 20 x 20 grid steps. After writing out a program title and table heading, the program reads in the required data. The variables NSTEPX and NSTEPY store the numbers of grid steps in the co-ordinate directions, while HOW is used for the channel depth to width ratio. The counter NBC determines the type of boundary conditions to be applied, as shown in Fig.

```
C   FINITE DIFFERENCE PROGRAM FOR FLOW ALONG A RECTANGULAR CHANNEL.
C
        DIMENSION  VELOC(21,21)
        WRITE(6,61)
   61   FORMAT(33H1FLOW ALONG A RECTANGULAR CHANNEL //
     1 64H    GRID    HOW  NBC    TOLER    ORELAX      F      ITER    NCA
     2LC)
    1   READ(5,51) NSTEPX,NSTEPY,HOW,NBC,TOLER,ORELAX
   51   FORMAT(2I3,F5.0,I4,2F10.0)
        IF(NSTEPX.EQ.0) STOP
        CALL  POISSN(VELOC,21,21,NSTEPX,NSTEPY,1.,HOW,NBC,TOLER,ORELAX,
     1                ITER,500)
C
C   INTEGRATE THE DIMENSIONLESS VELOCITY TO FIND THE SHAPE FACTOR
C   FOR THE FLOW RATE.
        IMAX=NSTEPX+1
        JMAX=NSTEPY+1
        DO 3 J=1,JMAX
        Q=0.
        DO 2 I=1,NSTEPX
    2   Q=Q+VELOC(I,J)+VELOC(I+1,J)
    3   VELOC(1,J)=Q
        Q=0.
        DO 4 J=1,NSTEPY
    4   Q=Q+VELOC(1,J)+VELOC(1,J+1)
        Q=Q*0.25*HOW/FLOAT(NSTEPX*NSTEPY)
        IF(NBC.EQ.2) Q=Q*0.25
        F=-12.*Q/HOW**3
C
C   FIND THE PRODUCT OF THE NUMBER OF ITERATION CYCLES AND THE NUMBER
C   OF INTERNAL GRID POINTS.
        NCALC=(NSTEPX-1)*(NSTEPY-1)*ITER
C
C   PRINT OUT THE RESULTS.
        WRITE(6,62) NSTEPX,NSTEPY,HOW,NBC,TOLER,ORELAX,F,ITER,NCALC
   62   FORMAT(1X,I2,3H X ,I2,F7.2,I4,E12.4,F7.2,F12.5,I6,I8)
        GO TO 1
        END
```

Fig. 6.10   Main program for finding velocities and rate of flow along a rectangular channel

6.9. TOLER and ORELAX store the convergence tolerance and over-relaxation factor required for the Gauss–Seidel process.

Having called POISSN to compute the dimensionless velocities, the main program then evaluates the definite integral defined in equation (6.55). The trapezoidal rule described in Section 4.1.1 is used for this purpose, by first finding the integrals along the grid lines parallel to the X axis.

$$\theta_j = \sum_{i=1}^{p} \tfrac{1}{2} h_X \left(\psi_{ij} + \psi_{i+1 j} \right)$$

$$\int_0^{H/W} \int_0^1 \psi \, dX \, dY = \sum_{j=1}^{q} \tfrac{1}{2} h_Y \left(\theta_j + \theta_{j+1} \right)$$

(6.56)

where h_X and p are the step length and number of steps in the X-direction, and h_Y and q are their counterparts in the Y-direction. The trapezoidal rule is adequate for present purposes because its truncation error (equation (4.5)) is proportional to a higher power of the step length than that of the finite difference method (equation (6.19)). The program variable Q is used to accumulate the sums involved in the integration process, and the variables equivalent to ψ_{1j} are used to store the intermediate integrals θ_j. The required shape factor is obtained according to equation (6.55), the value of ϕ having been selected as one.

The variable NCALC is used to store the product of the number of internal grid points and the number of Gauss–Seidel iterations, which provides a measure of the number of calculations performed. Finally, the program writes out the input data, shape factor, number of iterations and number of calculations.

6.6.3 The finite difference subprogram

While the subprogram named POISSN, shown in Fig. 6.11, could in principle be used to solve a wide range of problems governed by Poisson's equation, there are some features which are only appropriate for the present problem. The solution domain is assumed to be rectangular, the parameter ϕ (equation (6.53), or in general, equation (6.5)) is independent of position, and the boundary conditions are as defined in Fig. 6.9, according to the value of the argument NBC.

The first argument PSI stores the required solutions ψ_{ij}, while NXDIM and NYDIM are used for dimensioning purposes. NSTEPX and NSTEPY supply the numbers of grid steps in the co-ordinate directions. The argument PHI enters the value of ϕ, which is defined as one by the main program. The remaining arguments include SFACT, which enters the value of H/W, the convergence tolerance, over-relaxation factor, iteration counter and maximum number of cycles of Gauss–Seidel iteration (defined as 500 by the main program).

The variables IMAX and JMAX are used to store the maximum number of points in the two co-ordinate directions. Before starting the Gauss–Seidel

```
        SUBROUTINE  POISSN(PSI,NXDIM,NYDIM,NSTEPX,NSTEPY,PHI,SFACT,NBC,
      1                    TOLER,ORELAX,ITER,NCYCLE)
C
C   SUBROUTINE FOR SOLVING POISSONS EQUATION IN A RECTANGULAR DOMAIN.
C
        DIMENSION  PSI(NXDIM,NYDIM)
        IF(NSTEPX.LT.NXDIM.AND.NSTEPY.LT.NYDIM) GO TO 1
        WRITE(6,61)
   61   FORMAT(30HODIMENSIONS EXCEEDED IN POISSN)
        STOP
C
C   SET INITIAL PROFILE AND BOUNDARY CONDITIONS FOR FIXED PSI.
    1   IMAX=NSTEPX+1
        JMAX=NSTEPY+1
        DO 2 I=1,IMAX
        DO 2 J=1,JMAX
    2   PSI(I,J)=0.
C
C   SET UP ITERATION LOOP.
        HX=1./FLOAT(NSTEPX)
        HY=SFACT/FLOAT(NSTEPY)
        GAMEW=1./HX**2
        GAMNS=1./HY**2
        GAMO=2.*(GAMEW+GAMNS)
        DO 6 ITER=1,NCYCLE
        SUMPS=0.
        SUMDPS=0.
        DO 3 I=2,NSTEPX
        DO 3 J=2,NSTEPY
        DELPSI=(GAMEW*(PSI(I-1,J)+PSI(I+1,J))+GAMNS*(PSI(I,J-1)
      1       +PSI(I,J+1))-PHI)/GAMO-PSI(I,J)
        SUMDPS=SUMDPS+ABS(DELPSI)
        PSI(I,J)=PSI(I,J)+DELPSI*ORELAX
    3   SUMPS=SUMPS+ABS(PSI(I,J))
C
C   TEST FOR CONVERGENCE.
        ERROR=SUMDPS/SUMPS
        IF(ERROR.LT.TOLER) RETURN
C
C   APPLY DERIVATIVE BOUNDARY CONDITIONS.
        IF(NBC.NE.2) GO TO 6
        DO 4 I=1,NSTEPX
    4   PSI(I,1)=(4.*PSI(I,2)-PSI(I,3))/3.
        DO 5 J=1,NSTEPY
    5   PSI(1,J)=(4.*PSI(2,J)-PSI(3,J))/3.
    6   CONTINUE
C
C   NORMAL EXIT FROM ITERATION LOOP INDICATES FAILURE TO CONVERGE.
        ITER=NCYCLE
        WRITE(6,62) ITER
   62   FORMAT(21HONO CONVERGENCE AFTER,I5,7H CYCLES)
        RETURN
        END
```

Fig. 6.11   Subprogram for solving Poisson's equation in a rectangular domain

process, all the ψ_{ij} are set to zero. Apart from providing suitable initial values at internal points, this also serves to define the prescribed zero values on the boundaries. The variables GAMEW, GAMNS and GAMO are used to store the values of γ_E or γ_W, γ_N or γ_S, and γ_O respectively, as defined in equations (6.49).

Inside the iteration loop, the coding is very similar to that used in the general Gauss–Seidel subprogram described in Section 5.4.4. The variables SUMPS and SUMDPS are used to sum the magnitudes of the ψ_{ij} and their changes between successive cycles of iteration which are temporarily stored in DELPSI. Over-relaxation is only applied at internal points. Where appropriate, derivative boundary conditions are applied at the end of each iteration cycle, with the aid of expressions equivalent to equation (6.24).

6.6.4 Results Before studying the computed results for shape factor, it is necessary to establish both the optimum over-relaxation factors, and the appropriate tolerance levels. The former are determined empirically by finding the numbers of cycles of iteration required for convergence to a tolerance of 10^{-2}, for values of ω in the range $1 < \omega < 2$. Figure 6.12 shows the results for 10 x 10 step grids. The shape of the curves for both types of boundary conditions is typical of the effect of over-relaxation. The number of cycles decreases steadily to a minimum as ω is increased, and then increases rather more rapidly (for NBC = 1 and $\omega = 2$, the process failed to converge in 500 cycles). Consequently, if the optimum ω is not known, it is safer to under-estimate rather than to overestimate its value. The effect of over-relaxation is considerable: Fig. 6.12 shows that savings of up to 65 per cent on the number of cycles of iteration are obtainable. The optimum values of ω for 10 x 10 step grids are about 1·6 and 1·95 for the two types of boundary conditions. The number of grid points, and hence the number of linear equations, also affects the optimum value. Figure 6.13 indicates the relevant values for the three grids used.

While a tolerance of 10^{-2} is appropriate for examining the effect of over-relaxation, it is too large to give accurate results for shape factor. Table 6.2 shows the variation of the computed value of shape factor with the convergence

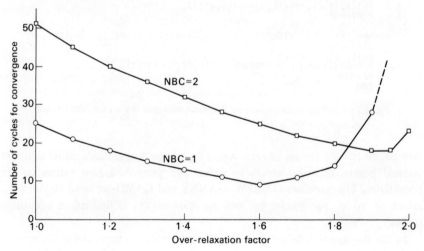

Fig. 6.12   Effect of over-relaxation on convergence to a tolerance of 10^{-2}, for 10 x 10 step grids

Table 6.2
Effect of convergence tolerance for 10 x 10 step grids

Convergence tolerance	Shape factors	
	NBC = 1	NBC = 2
10^{-2}	0·66367	0·60748
10^{-3}	0·66330	0·66879
10^{-4}	0·66216	0·67911
10^{-5}	0·66221	0·68014
10^{-6}	0·66222	0·68025

tolerance, for 10 x 10 step grids and using the relevant optimum over-relaxation factors. Two sources of error are involved: the truncation errors in the finite difference procedure, and the convergence errors in the Gauss–Seidel process. As indicated in Section 5.5.3, roundoff errors are not significant. If the convergence errors are negligible, the value of shape factor should not be affected by a change in the convergence tolerance. Any difference between the converged value of F and the exact solution is then due to truncation errors. Clearly, the convergence errors still have a small effect on F when the tolerance is as small as 10^{-6}, particularly for NBC = 2. Nevertheless, a tolerance of 10^{-6} is appropriate since the convergence errors are then small compared to the truncation errors.

Table 6.2 demonstrates a very important general point concerning testing for convergence, which was first raised in Section 3.1. A relative error involving the changes in the unknowns between successive cycles of iteration is compared with a prescribed tolerance. If the rate of convergence is slow, these changes diminish slowly as the number of iterations is increased. Consequently, the true relative error, which involves the sum of the changes over an infinite number of iterations, is much larger than the approximate error tested. Thus, in Table 6.2, the convergence errors are much larger than the corresponding tolerances.

Figure 6.13 shows the results obtained for the specified grid sizes and boundary conditions. A common convergence tolerance is employed, and the appropriate optimum over-relaxation factor is used in each case. The number of cycles of iteration for NBC = 2 (Fig. 6.9b) is about three times that required for NBC = 1 (Fig 6.9a) for the same size of grid. On the other hand, the computed

FLOW ALONG A RECTANGULAR CHANNEL

GRID	HOW	NBC	TOLER	ORELAX	F	ITER	NCALC
5 X 5	0.50	1	1.0000E-06	1.35	0.59802	15	240
5 X 5	0.50	2	1.0000E-06	1.92	0.66505	49	784
10 X 10	0.50	1	1.0000E-06	1.60	0.66222	28	2268
10 X 10	0.50	2	1.0000E-06	1.95	0.68025	100	8100
20 X 20	0.50	1	1.0000E-06	1.80	0.67993	58	20938
20 X 20	0.50	2	1.0000E-06	1.98	0.68452	188	67868

Fig. 6.13   Finite difference results for channel-flow problem

value of shape factor is much more accurate. As is to be expected, the results for NBC = 2 are very similar to those obtained for NBC = 1 using twice as many grid steps in each direction. The best parameter to use for comparing computational efficiencies is the number of calculations under the heading 'NCALC'. On this basis it is clear that, for a given accuracy of solution, the use of NBC = 2 saves over 60 per cent of the amount of arithmetic involved in analysing flow in the whole channel cross-section.

For the present problem, further improvements in accuracy can be achieved with the aid of the h^2 extrapolation process described in Section 6.4.4. The results for NBC = 2 and grid sizes of 10 x 10 and 20 x 20 are respectively 0·85 and 0·22 per cent in error compared to the accurate solution $F = 0·6860$. Using equation (6.45), an improved estimate can be obtained as

$$F = \tfrac{1}{3} (4 \times 0·68452 - 0·68025) = 0·68594$$

which is almost correct to four significant figures.

The relative advantages and disadvantages of direct elimination and iterative methods for solving simultaneous linear equations were discussed in Section 5.5. In particular, a criterion for the Gauss–Seidel method to be more efficient than gaussian elimination in terms of computing time was stated in equation (5.39). It is interesting to apply this test to the results shown in Fig. 6.13. As there are only five nonzero elements per row of the coefficient matrix, the criterion for preferring the Gauss–Seidel method can be expressed as

$$r = 15 \, m/n^2 < 1 \tag{6.57}$$

where m is the number of iterations and n the number of equations. Table 6.3 shows the variation of the relative efficiency parameter r with grid size and type of boundary conditions. With one exception, r is less than one, and the relative efficiency of the Gauss–Seidel process increases rapidly with the number of equations. As mentioned in Section 5.5, however, there are ways to improve the efficiency of direct methods of solution for sparse sets of equations.

Table 6.3
Variation of the relative efficiency parameter

Grid	NBC	m	n	r
5 x 5	1	15	16	0·88
5 x 5	2	49	16	2·9
10 x 10	1	28	81	0·064
10 x 10	2	100	81	0·23
20 x 20	1	58	361	0·0067
20 x 20	2	188	361	0·022

6.7   Propagation Problems

The distinction between equilibrium and propagation problems was discussed in Section 6.1, where it was shown that the latter are associated with either parabolic or hyperbolic partial differential equations. General methods for solving such equations are beyond the scope of this book (but are discussed by, for example, Crandall (1956)), and attention is confined to one particular method for solving simple parabolic equations. When applied to time-dependent thermal conduction problems it is generally known as the Schmidt method (see, for example, Schenck (1960)). If there is no heat generation in the solution domain, equation (6.12) becomes

$$\frac{\partial^2 T}{\partial x^2} = \frac{1}{\kappa} \frac{\partial T}{\partial t} \tag{6.58}$$

where $\kappa = k/\rho C_p$ is the thermal diffusivity of the medium concerned.

The following analysis is applicable to any parabolic equation similar in form to equation (6.58), provided the range of x is independent of t. The solution domain at any instant of time is as shown in Fig. 6.2. In order to avoid a subsequent confusion in notation, however, δx is used here for the constant step length. Following equation (6.18), the required space derivative at the point labelled i can be approximated by

$$\left(\frac{\partial^2 T}{\partial x^2}\right)_i \approx \frac{T_{i-1} - 2T_i + T_{i+1}}{(\delta x)^2} \tag{6.59}$$

where the T_i are the temperatures at time t. If at time $t + \delta t$, where δt is a small increment, the temperatures have changed to T_i', then the required time derivative at the same typical point i can be approximated by

$$\left(\frac{\partial T}{\partial t}\right)_i \approx \frac{T_i' - T_i}{\delta t} \tag{6.60}$$

Substitution of these expressions into equation (6.58) gives

$$T_i' = \frac{\kappa \, \delta t}{(\delta x)^2} (T_{i-1} + T_{i+1}) + T_i \left(1 - \frac{2\kappa \, \delta t}{(\delta x)^2}\right) \tag{6.61}$$

Now if the time increment is selected as

$$\delta t = \tfrac{1}{2}(\delta x)^2/\kappa \tag{6.62}$$

the new temperatures are obtained as the averages of those at the two adjacent points at the beginning of the time increment

$$T_i' = \tfrac{1}{2}(T_{i-1} + T_{i+1}) \tag{6.63}$$

This result can be applied to all internal points, the boundary temperatures being determined by the relevant conditions.

For example, the temperature of the left-hand boundary in Fig. 6.2 might be a prescribed function of time

$$T_1 = F(t), \quad T_1' = F(t + \delta t) \tag{6.64}$$

Another common type of boundary condition occurs when a solid medium forming the solution domain is exposed to a fluid medium and the surface heat transfer coefficient, h, is known. For example, if the right-hand boundary in Fig. 6.2 is of this type, the heat transfer rate, per unit surface area, to the fluid at a particular time is

$$q = h (T_n - T_\infty) = -k \left(\frac{\partial T}{\partial x} \right)_n \tag{6.65}$$

where T_∞ is the bulk temperature of the fluid. This equation can be used to define the boundary temperature with the aid of a finite difference approximation, equivalent to equation (6.23), for the derivative at the boundary

$$T_n = (\beta T_\infty + 4T_{n-1} - T_{n-2})/(3 + \beta) \tag{6.66}$$

where $\beta = 2h \, \delta x/k$. If the initial temperature distribution is known, equations (6.63), (6.64) and (6.66) can be used to determine the distribution at any subsequent time.

This method is of the *explicit* type, so-called because the space derivative is defined in terms of the known temperatures at the beginning of the time step. Such a method is unstable, with temperatures tending to infinite values, if the time step is too large. A step length defined according to equation (6.62) is acceptable from this point of view, although it may entail an excessively large number of steps in particular physical problems. The only way of increasing δt is to use an *implicit* method of solution which takes some account of the space derivative at the end of the current time step.

As with all finite difference methods, the solutions obtained are subject to significant errors. The truncation errors in equations (6.59) and (6.60) are of the order of $(\delta x)^2$ and δt respectively. In view of equation (6.62), however, δt is proportional to $(\delta x)^2$, and the accuracy is of the same order as the other finite difference methods considered in this chapter.

6.8   Case Study: Transient Heat Transfer in the Wall of a Furnace

The Schmidt method for time-dependent thermal conduction can be illustrated with the aid of a problem concerning the temperature distribution in the wall of a furnace.

6.8.1 Problem specification   The wall of furnace has a thickness of 0·2 m, and is composed of an insulating material having a thermal diffusivity of 5×10^{-4} $m^2 h^{-1}$. The inner surface of the wall is raised to a temperature of 500 K above ambient for 8 hours, and is assumed to be at ambient for the next 16 hours. The outer surface is exposed to the air and the surface heat transfer coefficient is estimated to be such that $k/h = 0·06$ m, where k is the thermal conductivity of the wall. Assuming that the wall is initially at ambient temperature, the temperature of its outer surface is to be computed as a function of time over the 24 hour period. The maximum value of this temperature and the time at which it occurs are to be found. The wall thickness is to be divided into 64 increments for the finite difference analysis.

6.8.2 Solution   Figure 6.14 shows a program for this problem. The arrays T and TNEW are used to store the temperatures at the beginning and end of the current time interval. The zero for these temperatures is taken to be at ambient conditions (T_∞ in the analysis). The arrays are dimensioned to allow for up to a hundred space increments through the wall thickness. The variables DIFFUS, KOH and L are used to store thermal diffusivity, the ratio k/h, and the wall thickness, while TDIFF stores the temperature difference across the wall when the furnace is in use. The time periods for which the inner wall surface are hot and cold are stored in HTIME and CTIME respectively, and the number of space increments and the time interval after which the temperatures are to be written out are stored in NSTEP and TINTW.

The computed values of δx and δt (equation (6.62)) are stored in DELTAX and DELTAT respectively. IFREQX is employed to store a quarter of the value of NSTEP (which should therefore be divisible by four), and is used subsequently for writing out temperatures. The variable N stores the number of points in the solution domain. After setting the initial temperature profile, the program determines the maximum value for the time step counter, JMAX. The variable TEMMAX, which is used to store the maximum temperature of the outer surface, is set to zero initially, and the value of β used in equation (6.66) is stored in BETA. The variable WTIME is used to store the next time at which temperatures are to be written out, and is therefore set equal to TINTW initially.

Inside the time step loop, the current time is stored in TIME and the inner surface temperature is defined. The new internal temperatures are then determined according to equation (6.63), and the outer surface temperature according to equation (6.66). If the new value of the latter temperature is greater than the stored maximum value then this new maximum is stored, and the corresponding time is stored in TIMMAX. The temperature profile, in terms of five equally spaced values, is written out when the current time just exceeds the value of WTIME, which is then increased by the value of TINTW. When the solution process is complete, the maximum outer surface temperature and corresponding time are written out.

```
C   PROGRAM FOR TRANSIENT TEMPERATURE DISTRIBUTION IN FURNACE WALL.
C
      DIMENSION  T(101),TNEW(101)
      REAL  KOH,L
      WRITE(6,61)
   61 FORMAT(51H1TRANSIENT TEMPERATURE DISTRIBUTION IN FURNACE WALL)
    1 READ(5,51) DIFFUS,KOH,L,TDIFF,HTIME,CTIME,NSTEP,TINTW
   51 FORMAT(F10.0,5F5.0,I5,F5.0)
      IF(NSTEP.LE.0.OR.NSTEP.GT.100) STOP
C
C   CALCULATE SPACE AND TIME INCREMENTS.
      DELTAX=L/FLOAT(NSTEP)
      IFREQX=NSTEP/4
      DELTAT=0.5*DELTAX**2/DIFFUS
      WRITE(6,62) DIFFUS,KOH,L,TDIFF,HTIME,CTIME,NSTEP,DELTAT
   62 FORMAT(30H THERMAL DIFFUSIVITY OF WALL =,E12.4,7H M**2/H /
     1  6H K/H =,F6.3,2H M,10X,16HWALL THICKNESS =,F6.3,2H M /
     2 25H TEMPERATURE DIFFERENCE =,F6.0,2H K /
     3 11H HOT TIME =,F5.1,2H H,10X,12H COLD TIME =,F5.1,2H H /
     4 29H NUMBER OF SPACE INCREMENTS =,I4,5X,15HTIME INTERVAL =,F6.3,
     5  2H H / 49H        HOURS     T(1)                      T(N)  )
C
C   SET INITIAL TEMPERATURES AND CONTROL PARAMETERS.
      N=NSTEP+1
      T(1)=TDIFF
      DO 2 I=2,N
    2 T(I)=0.
      JMAX=(HTIME+CTIME)/DELTAT+1.5
      TEMMAX=0.
      BETA=2.*DELTAX/KOH
      WTIME=TINTW
C
C   SET UP TIME INCREMENT LOOP.
      DO 6 J=1,JMAX
      TIME=DELTAT*FLOAT(J)
      TNEW(1)=TDIFF
      IF(TIME.GT.HTIME) TNEW(1)=0.
C
C   CALCULATE NEW TEMPERATURES.
      DO 3 I=2,NSTEP
    3 TNEW(I)=0.5*(T(I-1)+T(I+1))
      TNEW(N)=(4.*TNEW(N-1)-TNEW(N-2))/(3.+BETA)
C
C   STORE AS OLD TEMPERATURES FOR NEXT TIME INCREMENT.
      DO 4 I=1,N
    4 T(I)=TNEW(I)
C
C   TEST FOR MAXIMUM SURFACE TEMPERATURE.
      IF(TNEW(N).LE.TEMMAX) GO TO 5
      TEMMAX=TNEW(N)
      TIMMAX=TIME
C
C   OUTPUT RESULTS AT REQUIRED TIME INTERVALS.
    5 IF(TIME.LT.WTIME) GO TO 6
      WTIME=WTIME+TINTW
      WRITE(6,63) TIME,(T(I),I=1,N,IFREQX)
   63 FORMAT(1X,F8.2,5F8.0)
    6 CONTINUE
      WRITE(6,64) TEMMAX,TIMMAX
   64 FORMAT(36H MAXIMUM OUTER SURFACE TEMPERATURE =,F5.0,6H K, AT,F5.1,
     1  2H H )
      GO TO 1
      END
```

Fig. 6.14   Program for finding transient temperature distributions in the wall of a furnace

6.8.3 Results   Figure 6.15 shows the results for the specified problem, for
TINTW = 2 hours (not exactly divisible by δt). The maximum temperature of
the outer surface of the furnace wall is 42 K above ambient, which is achieved
nearly 7 hours after the temperature of the inner surface is reduced to ambient.
This demonstrates the 'thermal wave' phenomenon where a temperature peak
moves slowly through a conducting medium.

```
TRANSIENT TEMPERATURE DISTRIBUTION IN FURNACE WALL
THERMAL DIFFUSIVITY OF WALL =  0.5000E-03 M**2/H
K/H = 0.060 M              WALL THICKNESS = 0.200 M
TEMPERATURE DIFFERENCE =  500. K
HOT TIME =  8.0 H                COLD TIME = 16.0 H
NUMBER OF SPACE INCREMENTS =  64        TIME INTERVAL = 0.010 H
   HOURS     T(1)                                       T(N)
    2.00     500.     132.      13.       0.       0.
    4.00     500.     215.      57.       9.       1.
    6.01     500.     259.      98.      27.       7.
    8.01       0.     288.     132.      48.      17.
   10.00       0.     177.     147.      68.      28.
   12.00       0.     110.     125.      79.      38.
   14.00       0.      77.     102.      78.      42.
   16.01       0.      59.      84.      72.      42.
   18.01       0.      47.      70.      65.      39.
   20.00       0.      38.      59.      57.      35.
   22.00       0.      32.      50.      49.      30.
   24.00       0.      27.      43.      42.      26.
MAXIMUM OUTER SURFACE TEMPERATURE =  42. K, AT 14.7 H
```

Fig. 6.15   Results from program for transient temperature distributions

7 Structural Analysis and Finite Element Methods

The analysis of engineering structures is an important subject, both in its own right and because it leads naturally into so-called finite element methods. Such methods find applications in a very wide range of engineering problems, including the type of continuum mechanics problems discussed in Chapter 6. In this chapter, the analysis of simple structures is considered, and in order to emphasise the link with finite element methods the members of such structures are referred to as elements. Various finite element methods are discussed, and compared with finite difference methods, and one relatively simple example is studied in detail.

7.1 Pin-Jointed Structures

Among the simplest types of structures are frameworks consisting of relatively long thin elements (members) pin-jointed at their ends. The elements may therefore be subjected to tension or compression, but not to bending. Figure 7.2, which illustrates the problems studied in Section 7.2, shows some examples of pin-jointed structures. Although attention is confined here to plane frameworks, the method of analysis can be extended to space frames.

The purpose of analysing pin-jointed structures is to determine both the forces in the elements and the displacements of the joints or *nodes*, under specified conditions of loading. When such problems are solved by hand, the distinction between *statically determinate* and *statically indeterminate* structures is an important one. As the forces in the elements of statically determinate structures can be found without reference to the displacements, such structures are easier to analyse. The method presented here, however, can be applied to any pin-jointed structure. The basis of the method is to treat the displacements of the nodes as the unknowns to be found. An alternative approach is to treat the forces in the elements as the unknowns.

The first stage in the analysis is to examine the behaviour of individual

144

elements. Figure 7.1 shows a typical element lying in the x-y plane, inclined at an angle θ to the x-direction. All the elements are numbeied, and the number of this typical element is m. Its length, cross-sectional area and Young's modulus are L_m, A_m and E_m respectively. The numbers of the associated nodal points are i and j. When the structure is loaded, the displacements of the typical point i in the co-ordinate directions are u_i and v_i, and the corresponding force components applied to the element are U_i and V_i.

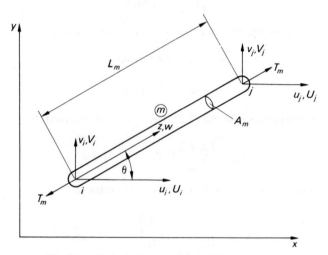

Fig. 7.1   Typical element of a pin-jointed structure

It is convenient to define a local co-ordinate, z, along the element, and a corresponding displacement component w. As the cross-sectional area of the element is uniform, the variation of this displacement along the element is given by

$$w = C_1 + C_2 z \qquad (7.1)$$

where

$$C_1 = w_i, \quad C_2 = (w_j - w_i)/L_m \qquad (7.2)$$

Now the strain in the element is

$$\frac{dw}{dz} = C_2 = (w_j - w_i)/L_m \qquad (7.3)$$

and the (tensile) force is

$$T_m = E_m A_m \frac{dw}{dz} = \frac{E_m A_m}{L_m}(w_j - w_i) \qquad (7.4)$$

The forces and displacements must now be expressed in terms of the components in the directions of the 'global' co-ordinates, x and y. From Fig. 7.1, the force components applied to the element are related to T_m as follows

$$f_m = \begin{bmatrix} U_i \\ V_i \\ U_j \\ V_j \end{bmatrix} = \begin{bmatrix} -\cos\theta \\ -\sin\theta \\ \cos\theta \\ \sin\theta \end{bmatrix} T_m \tag{7.5}$$

and the relationships between the displacement components are

$$w_i = u_i \cos\theta + v_i \sin\theta$$
$$w_j = u_j \cos\theta + v_j \sin\theta \tag{7.6}$$

Hence, using equations (7.4) and (7.6), the force in the element is

$$T_m = \frac{E_m A_m}{L_m} [-c \;\; -s \;\; c \;\; s] \, d_m \tag{7.7}$$

where $c \equiv \cos\theta$, $s \equiv \sin\theta$, and the displacement vector is

$$d_m = \begin{bmatrix} u_i \\ v_i \\ u_j \\ v_j \end{bmatrix} \tag{7.8}$$

Hence, substituting T_m from equation (7.7) into equation (7.5), the relationship between the forces and displacements is

$$f_m = k_m d_m \tag{7.9}$$

where the element stiffness matrix, k_m, is given by

$$k_m = \frac{E_m A_m}{L_m} \begin{bmatrix} -c \\ -s \\ c \\ s \end{bmatrix} [-c \;\; -s \;\; c \;\; s] = \frac{E_m A_m}{L_m} \begin{bmatrix} c^2 & cs & -c^2 & -cs \\ cs & s^2 & -cs & -s^2 \\ -c^2 & -cs & c^2 & cs \\ -cs & -s^2 & cs & s^2 \end{bmatrix} \tag{7.10}$$

The next stage in the analysis is to add together the force–displacement characteristics of the individual elements to determine the behaviour of the overall structure. Since the elements are connected at the nodes, the displacement of a particular node is the same for every element connected to it.

Also, the conditions for equilibrium of the structure may be expressed in words and symbols as

$$\begin{pmatrix} \text{externally applied} \\ \text{forces at the nodes} \end{pmatrix} = \sum \begin{pmatrix} \text{forces on the elements} \\ \text{at these nodes} \end{pmatrix}$$

$$F = \Sigma f_m = \Sigma k_m d_m = K D \qquad (7.11)$$

where K is the overall stiffness matrix, and the vectors F and D contain the externally applied forces and the corresponding displacements. For example, if there are n nodes, D contains $u_1, v_1, u_2, v_2, \ldots, u_n, v_n$ in this order. The summation in equation (7.11) is for all the elements in the structure, and the coefficients of the overall stiffness matrix are assembled from the coefficients of the stiffness matrices of the individual elements. The matrix K is generally sparse because the number of elements connected to a particular nodal point is usually small.

Let K_{pq} and k_{rs} be typical coefficients of the overall and element stiffness matrices respectively, where p and q lie in the range 1 to $2n$, while r and s lie in the range 1 to 4. The subscripts p and r are row numbers, while q and s are column numbers. Now k_{rs} can be interpreted as the force that must be applied to the typical element at the node and in the direction corresponding to the rth coefficient of the element force vector shown in equation (7.5), to cause a unit displacement at the node and in the direction corresponding to the sth coefficient of the element displacement vector shown in equation (7.8). A similar interpretation can be applied to K_{pq} in terms of the pth coefficient of the vector of externally applied forces and the qth coefficient of D. The process of assembling the overall stiffness coefficients takes the form of

$$K_{pq} = \Sigma k_{rs} \qquad (7.12)$$

where the row and column numbers are equivalent. For example, if $r < 3$ the nodal point number concerned is i, and the equivalent value of p is $2(i-1) + r$. Similarly, if $s \geqslant 3$ the nodal point number concerned is j, and the equivalent value of q is $2(j-1) + (s-2)$. Assembly is complete when the sixteen coefficients of each and every element stiffness matrix have been added to the relevant overall stiffness coefficients.

Before equations (7.11) can be solved for the displacements, the appropriate restraints must be applied. The position of at least one node of the structure must be fixed. Another type of condition which may also be appropriate is where a particular node is allowed to move in one direction only. For example, a bridge structure might be pin-jointed to its foundations at one end, and free to slide horizontally at the other. These conditions can be accommodated by modifying the equations, which are then solved to give

$$D = K^{-1} F \qquad (7.13)$$

As the typical element stiffness matrix displayed in equation (7.10), and hence the overall stiffness matrix, are not diagonally dominant, the gaussian elimination method of solution is appropriate. Once the displacements have been found, the forces in the individual elements can be obtained with the aid of equation (7.7).

While a pin-jointed structure can be statically determinate or statically indeterminate, it is also possible for it to be a mechanism. If the present method of solution is applied to such a structure, a singular overall stiffness matrix is obtained, implying that at least some of the displacements are infinite. If the structure is not a mechanism, it is a simple matter to determine whether it is statically determinate. Since there are $2n$ equilibrium equations for the n nodes, then if there are e elements and t external reaction components applied at the restrained nodes, the forces in the elements could be found without reference to the displacements of the nodes if

$$e + t = 2n \qquad\qquad (7.14)$$

For every fixed node there are two reaction components, and for a node restrained to move freely in a particular direction there is one such component. If $e + t > 2n$, the structure is statically indeterminate.

7.2   Case Study: Displacements of Plane Pin-Jointed Frameworks

Figure 7.2a shows a typical pin-jointed framework with equal vertical loads at three of its nodes. There are ten elements joined at seven nodes, two of which are fixed. Since the structure is not a mechanism and equation (7.14) is satisfied, it is therefore statically determinate. If, for example, the element numbered 3 is relocated as shown in Fig. 7.2b, a mechanism is obtained, although equation (7.14) is still satisfied. Figure 7.2c, on the other hand, shows the first structure with two elements added to make it statically indeterminate.

7.2.1 Problem specification In the structures shown in Fig. 7.2, all the elements have cross-sectional areas of 12.5 cm^2, and lengths of either 3 or $3\sqrt{2}$ m. The Young's modulus of the steel used is 208 GN m^{-2}, and the three vertical loads are each 50 kN. The displacements of the nodal points and the forces in the elements are to be computed for each of these structures. The program should detect whether a particular structure is a mechanism, statically determinate or statically indeterminate.

7.2.2 Solution Figure 7.3 shows a main program for analysing plane pin-jointed structures, with the aid of the gaussian elimination subprogram ELIMIN shown in Fig. 5.3. The DIMENSION statement allows for structures involving up

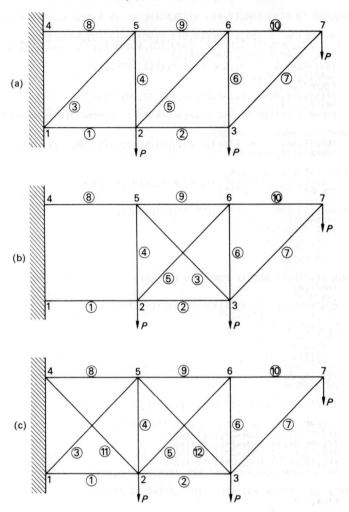

Fig. 7.2   Some pin-jointed frameworks

to twenty elements and twenty nodal points to be analysed. The arrays F, STIFF and D are used to store the externally applied nodal point forces, the coefficients of the overall stiffness matrix, and the displacements of the nodal points respectively. Note that STIFF is allowed an extra column so that it can be extended to include the nodal point forces, as required by ELIMIN. While F and D are in single subscript form, for some purposes double subscripts are more convenient: one for the nodal point number and one for the direction. Hence, arrays LOAD and UV are introduced and refer to the same storage registers as F and D by virtue of the EQUIVALENCE statement. The variables UV(1, I) and UV(2, I), for example, store the values of u_i and v_i respectively, where i is the

```
C   PROGRAM TO FIND FORCES AND DISPLACEMENTS IN A PIN-JOINTED STRUCTURE.
C
        DIMENSION F(40),STIFF(40,41),D(40),UV(2,20),EFORCE(20),X(20),
      1 Y(20),NREST(20),NPI(20),NPJ(20),A(20),E(20),C(20),S(20),VECT(4)
        REAL  LOAD(2,20),L(20),K(4,4)
        EQUIVALENCE  (F(1),LOAD(1,1)),(D(1),UV(1,1))
    1   READ(5,51) NEL,NNP
   51   FORMAT(2I5)
        IF(NEL.LE.0.OR.NEL.GT.20.OR.NNP.GT.20) STOP
        WRITE(6,61)
   61   FORMAT(52H1FORCES AND DISPLACEMENTS IN A PIN-JOINTED STRUCTURE)
C
C   INPUT THE NODAL POINT DATA.
        READ(5,52) (I,NREST(I),X(I),Y(I),LOAD(1,I),LOAD(2,I),N=1,NNP)
   52   FORMAT(2I5,4F10.0)
C
C   INPUT THE ELEMENT DATA.
        READ(5,53) (M,NPI(M),NPJ(M),A(M),E(M),N=1,NEL)
   53   FORMAT(3I5,F10.0,E12.4)
C
C   PREPARE TO SUM THE STIFFNESS COEFFICIENTS.
        NEQN=2*NNP
        DO 2 IROW=1,NEQN
        DO 2 ICOL=1,NEQN
    2   STIFF(IROW,ICOL)=0.
        DO 3 M=1,NEL
C
C   FORM STIFFNESS MATRIX FOR EACH ELEMENT.
        I=NPI(M)
        J=NPJ(M)
        L(M)=SQRT((X(J)-X(I))**2+(Y(J)-Y(I))**2)
        C(M)=(X(J)-X(I))/L(M)
        S(M)=(Y(J)-Y(I))/L(M)
        FACT=E(M)*A(M)/L(M)
        VECT(1)=-C(M)
        VECT(2)=-S(M)
        VECT(3)=C(M)
        VECT(4)=S(M)
        DO 3 IRE=1,4
        DO 3 ICE=1,4
        K(IRE,ICE)=FACT*VECT(IRE)*VECT(ICE)
C
C   ADD ELEMENT STIFFNESS TO OVERALL STIFFNESS.
        IF(IRE.LT.3) IROW=2*(I-1)+IRE
        IF(IRE.GE.3) IROW=2*(J-1)+IRE-2
        IF(ICE.LT.3) ICOL=2*(I-1)+ICE
        IF(ICE.GE.3) ICOL=2*(J-1)+ICE-2
    3   STIFF(IROW,ICOL)=STIFF(IROW,ICOL)+K(IRE,ICE)
C
C   APPLY RESTRAINTS AND SOLVE THE LINEAR EQUATIONS.
        NREACT=0
        DO 6 I=1,NNP
        IF(NREST(I).EQ.0) GO TO 6
        NREACT=NREACT+2
        DO 5 N=1,2
        IROW=2*(I-1)+N
        DO 4 ICOL=1,NEQN
    4   STIFF(IROW,ICOL)=0.
        STIFF(IROW,IROW)=1.
    5   LOAD(N,I)=0.
    6   CONTINUE
C
C   EXTEND THE OVERALL STIFFNESS MATRIX TO INCLUDE THE FORCE VECTOR.
        DO 7 IROW=1,NEQN
    7   STIFF(IROW,NEQN+1)=F(IROW)
        CALL  ELIMIN(STIFF,D,NEQN,40,41,DET,RATIO)
        IF(DET.NE.0.) GO TO 8
        WRITE(6,62)
   62   FORMAT(29H0THE STRUCTURE IS A MECHANISM)
        GO TO 1
```

Fig. 7.3   Main program for analysing plane pin-jointed structures

```
C
C  FIND THE FORCES IN THE ELEMENTS.
8      DO 9 M=1,NEL
       I=NPI(M)
       J=NPJ(M)
       FACT=E(M)*A(M)/L(M)
9      EFORCE(M)=FACT*(-C(M)*UV(1,I)-S(M)*UV(2,I)+C(M)*UV(1,J)
      1                  +S(M)*UV(2,J))
C
C  OUTPUT THE RESULTS.
       WRITE(6,63) (M,NPI(M),NPJ(M),A(M),L(M),E(M),EFORCE(M),M=1,NEL)
63     FORMAT(64H0  ELEM    I    J     AREA        LENGTH         E
      1FORCE    /  (1X,3I5,4E12.4))
       WRITE(6,64) (I,NREST(I),X(I),Y(I),(LOAD(N,I),UV(N,I),N=1,2),
      1  I=1,NNP)
64     FORMAT(69H0NODE REST  X        Y      FX        U          FY
      1    V    /  (1X,2I3,2F7.1,4E12.4))
       IF(NEL+NREACT.EQ.NEQN) WRITE(6,65)
65     FORMAT(40H0THE STRUCTURE IS STATICALLY DETERMINATE)
       IF(NEL+NREACT.GT.NEQN) WRITE(6,66)
66     FORMAT(42H0THE STRUCTURE IS STATICALLY INDETERMINATE)
       GO TO 1
       END
```

Fig. 7.3 Continued

value of the counter I. The components of an array with more than one subscript are stored in an order such that the first subscript varies most rapidly.

The array EFORCE is used to store the tensile forces, T_m, in the elements, while X and Y store the co-ordinates of the nodal points. NREST stores integer numbers which define the type of restraint conditions applied to the nodal points. With the program in its present form, a zero value of NREST(I) means that the point whose number is given by the value of I is unrestrained, while any nonzero value means that the point is fixed. The program could be extended to include additional conditions such as displacement in a prescribed direction. The arrays NPI and NPJ are used to store the numbers of the nodal points (i and j for the typical element shown in Fig. 7.1) at the ends of the elements, while A, E and L store the element cross-sectional areas, Young's moduli and lengths. The cosines and sines of the angles of slope of the elements are stored in arrays C and S, and VECT is used for the vector of trigonometrical functions defined in equation (7.5). Finally, the array K is used to store the coefficients of the element stiffness matrices.

Other variables used in the program include NEL, NNP and NEQN for the number of elements, nodal points and equations respectively, while the counters I and J are used for nodal point numbers, and M for element numbers. IROW and ICOL are used for row and column numbers in the overall stiffness matrix, while IRE and ICE serve the same purpose for the element stiffness matrices (p, q, r and s are the equivalent counters in equation (7.12)).

After the input data have been read in, the coefficients of the overall stiffness matrix are set to zero in preparation for the assembly process. Then, for each element in turn, the numbers of the associated nodal points are assigned to I and J, and the coefficients of the element stiffness matrix are computed according to equation (7.10). Each of these coefficients is added to the appropriate coefficient of the overall stiffness matrix, as indicated by equation (7.12).

The restraint conditions are applied according to the numbers stored in NREST. If a nonzero value is detected for a particular point numbered i, the position of that point is fixed by altering the corresponding equations to the forms $u_i = 0$ and $v_i = 0$. the value of NREACT, which stores the number of external reaction components applied at restrained nodes, is increased by two when such a fixed point is encountered.

In order that use can be made of subprogram ELIMIN, the overall stiffness matrix is extended to include the externally applied forces. If the overall stiffness matrix is singular (argument DET returned from ELIMIN as zero) the structure is a mechanism. The tensile forces in the elements are found with the aid of equation (7.7), and the results are written out. Finally, the structure is tested for statical determinacy according to equation (7.14).

There are a number of additions and improvements that could be made to the present program. Apart from the provision for further restraint conditions already mentioned, it is possible to allow for initial lack-of-fit of the elements in a statically-indeterminate structure. The effect of strains caused by changes of temperature can also be included. The program could be improved by incorporating much more detailed checks on the validity of the data read in.

7.2.3 Results   Figure 7.4 shows the results obtained for the structure shown in Fig. 7.2a. For each element, the corresponding nodal point numbers, dimensions, modulus and tensile force are printed. Then for each nodal point, the restraint condition number, co-ordinates, and applied force components and corresponding displacements are printed. The program can be tested by checking these results manually. For example, it can be checked that equilibrium is satisfied for the forces applied to each nodal point.

```
FORCES AND DISPLACEMENTS IN A PIN-JOINTED STRUCTURE
```

ELEM	I	J	AREA	LENGTH	E	FORCE
1	1	2	0.1250E-02	0.3000E+01	0.2080E+12	-0.1500E+06
2	2	3	0.1250E-02	0.3000E+01	0.2080E+12	-0.5000E+05
3	1	5	0.1250E-02	0.4243E+01	0.2080E+12	-0.2121E+06
4	2	5	0.1250E-02	0.3000E+01	0.2080E+12	0.1500E+06
5	2	6	0.1250E-02	0.4243E+01	0.2080E+12	-0.1414E+06
6	3	6	0.1250E-02	0.3000E+01	0.2080E+12	0.1000E+06
7	3	7	0.1250E-02	0.4243E+01	0.2080E+12	-0.7071E+05
8	4	5	0.1250E-02	0.3000E+01	0.2080E+12	0.3000E+06
9	5	6	0.1250E-02	0.3000E+01	0.2080E+12	0.1500E+06
10	6	7	0.1250E-02	0.3000E+01	0.2080E+12	0.5000E+05

NODE	REST	X	Y	FX	U	FY	V
1	1	0.0	0.0	0.E+00	-0.6448E-15	0.E+00	0.1429E-14
2	0	3.0	0.0	0.E+00	-0.1731E-02	-0.5000E+05	-0.1009E-01
3	0	6.0	0.0	0.E+00	-0.2308E-02	-0.5000E+05	-0.2143E-01
4	1	0.0	3.0	3.0	0.E+00	-0.5591E-15	0.E+00	0.E+00
5	0	3.0	3.0	3.0	0.E+00	0.3462E-02	0.E+00	-0.8357E-02
6	0	6.0	3.0	3.0	0.E+00	0.5192E-02	0.E+00	-0.2027E-01
7	0	9.0	3.0	3.0	0.E+00	0.5769E-02	-0.5000E+05	-0.3114E-01

```
THE STRUCTURE IS STATICALLY DETERMINATE
```

Fig. 7.4   Results for structure shown in Fig. 7.2a

If the data for the structure shown in Fig. 7.2b are supplied to the program, the only result is the message 'THE STRUCTURE IS A MECHANISM'. Figure 7.5 shows the results obtained for the structure shown in Fig. 7.2c. It is interesting to compare these results with those of Fig. 7.4 to find the effect of adding two extra elements and making the structure statically indeterminate. For example, the vertical displacement of the node furthest from the support points is reduced from 0·031 m to 0·023 m.

```
FORCES AND DISPLACEMENTS IN A PIN-JOINTED STRUCTURE

ELEM    I    J      AREA        LENGTH          E            FORCE
  1     1    2   0.1250E-02   0.3000E+01   0.2080E+12   -0.2270E+06
  2     2    3   0.1250E-02   0.3000E+01   0.2080E+12   -0.1076E+06
  3     1    5   0.1250E-02   0.4243E+01   0.2080E+12   -0.1032E+06
  4     2    5   0.1250E-02   0.3000E+01   0.2080E+12    0.1543E+05
  5     2    6   0.1250E-02   0.4243E+01   0.2080E+12   -0.6002E+05
  6     3    6   0.1250E-02   0.3000E+01   0.2080E+12    0.4244E+05
  7     3    7   0.1250E-02   0.4243E+01   0.2080E+12   -0.7071E+05
  8     4    5   0.1250E-02   0.3000E+01   0.2080E+12    0.2230E+06
  9     5    6   0.1250E-02   0.3000E+01   0.2080E+12    0.9244E+05
 10     6    7   0.1250E-02   0.3000E+01   0.2080E+12    0.5000E+05
 11     2    4   0.1250E-02   0.4243E+01   0.2080E+12    0.1089E+06
 12     3    5   0.1250E-02   0.4243E+01   0.2080E+12    0.8140E+05

NODE REST   X      Y      FX          U            FY            V
  1    1   0.0    0.0   0.E+00    0.6018E-15    0.E+00      0.1520E-15
  2    0   3.0    0.0   0.E+00   -0.2619E-02   -0.5000E+05  -0.5133E-02
  3    0   6.0    0.0   0.E+00   -0.3860E-02   -0.5000E+05  -0.1327E-01
  4    1   0.0    3.0   0.E+00   -0.7647E-15    0.E+00      -0.1296E-14
  5    0   3.0    3.0   0.E+00    0.2573E-02    0.E+00      -0.4955E-02
  6    0   6.0    3.0   0.E+00    0.3640E-02    0.E+00      -0.1278E-01
  7    0   9.0    3.0   0.E+00    0.4216E-02   -0.5000E+05  -0.2298E-01

THE STRUCTURE IS STATICALLY INDETERMINATE
```

Fig. 7.5   Results for structure shown in Fig. 7.2c

7.3 Rigid-Jointed Structures

Many structures involve elements which are rigidly jointed. In addition to acting as simple struts and stays as in pin-jointed structures, these elements can also take bending loads. In many cases, only the contributions of the bending stiffnesses to the overall stiffness need be considered.

Rigid-jointed structures can be analysed by a method which is very similar in principle to that used for pin-jointed structures. If the elements are of the form shown in Fig. 7.1, then in addition to forces and displacements, the bending moments and rotations at the nodal points must also be considered. The analysis of a typical element (see, for example, Desai and Abel (1972)) yields a relationship analogous to equation (7.9), where f_m and d_m now include moments and rotations respectively. The number of equations for equilibrium of the overall structure can be increased to include the conditions for equilibrium of moments at the nodal points.

An important distinction between pin and rigid-jointed structures is that the latter are continuous. For example, the method of analysis just outlined can be applied to a cantilevered beam which is divided into a finite number of elements, not necessarily of equal length. Even if the beam is curved, or its cross-section changes, its behaviour as a structure can be approximated by treating it as a series of simple straight elements. The same method can be extended to the analysis of plates and shells.

7.4   Finite Element Methods

The above discussion of the analysis of rigid-jointed structures leads naturally into the consideration of finite element methods, although these are not confined to structural applications. A structure or solid medium can be divided up into a finite number of simple elements. The flow of fluid media can also be analysed by the same approach, although the elements are then spatial rather than material subregions of the continuum.

The detailed study of finite element methods is beyond the scope of this book, and the interested reader should consult textbooks such as Desai and Abel (1972) or Zienkiewicz (1971). The purpose of the following discussion is to briefly outline the application of finite element methods to continuum problems, and to compare them with finite difference methods. Although there are numerous types of elements, they may be categorised as either one, two or three-dimensional.

7.4.1 One-dimensional elements The analysis of pin-jointed structures presented in Section 7.1 provides an example of the use of one-dimensional finite elements. In the derivation that follows, and in the case study considered in Section 7.5, there is a very close similarity between this structural analysis and the finite element analysis of one-dimensional continuum problems.

While finite difference methods are applied to particular types of differential equations, finite element methods are often applied to particular physical problems and the analyses evolved with the aid of direct physical arguments. The laminar flow of a viscous newtonian fluid along a rectangular channel has already been studied in Sections 3.2 and 6.6. If the channel is very wide compared to its depth ($W \gg H$ in Fig. 3.1) it is only necessary to consider velocity variations through the depth of the channel, and the following analysis involving one-dimensional finite elements can be used.

Figure 7.6a shows the form of velocity profile in the channel. The gap between the flow boundaries is divided into a number of one-dimensional spatial elements, which are not necessarily of equal length. While the elements are joined at discrete nodes as shown, it is convenient to think of them as having unit dimensions in the z and x-directions. There are n nodes and $n - 1$ elements.

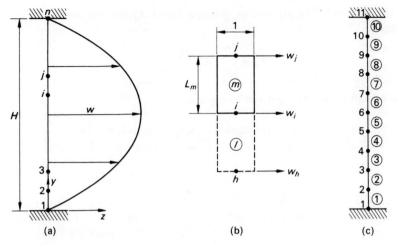

Fig. 7.6   One-dimensional finite element analysis of fluid flow. (a) Velocity profile;
(b) Typical element; (c) Arrangement of elements used in Section 7.5

Figure 7.6b shows a typical element, numbered m, of length L_m, and involving the nodal points numbered i and j. The velocities at these points are w_i and w_j respectively. Assuming a linear variation of velocity over the element

$$w = C_1 + C_2 y \tag{7.15}$$

where

$$C_1 = w_i, \quad C_2 = (w_j - w_i)/L_m \tag{7.16}$$

As the fluid is newtonian, with a viscosity μ, the shear stress acting on the element is given by

$$\tau = \mu \frac{dw}{dy} = \mu(w_j - w_i)/L_m \tag{7.17}$$

The viscous forces W_i and W_j acting on the element at its nodes, due to the stresses on the lower and upper surfaces, are

$$\begin{bmatrix} W_i \\ W_j \end{bmatrix} = \begin{bmatrix} -\tau \\ \tau \end{bmatrix} = \frac{\mu}{L_m} \begin{bmatrix} 1 & -1 \\ -1 & 1 \end{bmatrix} \begin{bmatrix} w_i \\ w_j \end{bmatrix} \tag{7.18}$$

That is

$$W_m = k_m w_m \tag{7.19}$$

where k_m is the viscous stiffness matrix. The effect of the pressure gradient, P_z, is to apply a uniformly distributed force in the z-direction of $-P_z$ per unit area to the element. The equivalent pressure forces applied to the element at its nodes are

$$f_m = \begin{bmatrix} P_i \\ P_j \end{bmatrix} = -\frac{P_z L_m}{2} \begin{bmatrix} 1 \\ 1 \end{bmatrix} \tag{7.20}$$

The equilibrium conditions can be expressed as

$$\sum \begin{pmatrix} \text{pressure forces applied} \\ \text{at the nodes} \end{pmatrix} = \sum \begin{pmatrix} \text{viscous forces on the} \\ \text{elements at these nodes} \end{pmatrix}$$

$$F = \sum f_m = \sum k_m w_m = K w \tag{7.21}$$

where K is the overall viscous stiffness matrix, and the vectors F and w contain the applied pressure forces and corresponding velocities. For example, w contains $w_1, w_2, w_3, \ldots, w_n$. Let K_{pq} and k_{rs} be typical coefficients of the overall and element viscous stiffness matrices, and let F_p and f_r be typical coefficients of the overall and element pressure force vectors. The values of p and q lie in the range 1 to n, while those of r and s lie in the range 1 to 2. The process of assembling stiffnesses and pressure forces takes the form of

$$K_{pq} = \sum k_{rs}, \quad F_p = \sum f_r \tag{7.22}$$

When the typical element shown in Fig. 7.6b is involved, p takes the value of i or j according to whether the value of r is 1 or 2, and similarly for q according to the value of s.

Consider, for example, the forces at the typical nodal point i in Fig. 7.6b, and assume that the other adjacent element and nodal point are numbered respectively l and h as shown. The equilibrium condition is

$$\mu(-w_h + w_i)/L_l + \mu(w_i - w_j)/L_m = -\tfrac{1}{2}P_z L_l - \tfrac{1}{2}P_z L_m \tag{7.23}$$

and if $L_l = L_m = L$, this becomes

$$w_h - 2w_i + w_j = L^2 P_z/\mu \tag{7.24}$$

which is identical to the finite difference equation (6.20) when applied to the one-dimensional form of the relevant differential equation, equation (6.4).

Boundary conditions $w_1 = w_n = 0$ can be applied by making appropriate modifications to the first and last rows of K and F, and equations (7.21) can then be solved for the w_i. As the nodal points are numbered in sequence from 1 to n (Fig. 7.6a) the overall stiffness matrix is tridiagonal. The solution of this problem is considered in Section 7.5.

7.4.2 Two-dimensional elements
The above analysis of flow using one-dimensional finite elements can be extended to the problem of flow in a channel

of finite width, with the aid of two-dimensional elements. For example, the cross-section of the channel can be divided into triangular elements, of the form shown in Fig. 7.7. As there are three nodal points, one at each corner, associated with each element, a linear variation of velocity over the element is assumed

$$w = C_1 + C_2 x + C_3 y \tag{7.25}$$

The three constants can be expressed in terms of the velocities at the three nodal points. The subsequent analysis is similar to that for one-dimensional elements, with equilibrium conditions for the forces at the nodal points being derived.

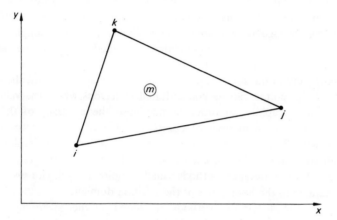

Fig. 7.7 Triangular two-dimensional finite element

This direct equilibrium analysis can only be applied to simple elements in which stresses are constant. In general, two-dimensional elements may have more than three sides, and may have nodes in addition to those at the corners. For example, the element shown in Fig. 7.7 could have extra nodes at the mid points of its sides. A quadratic variation of velocity over the element would then be assumed, involving six constants expressible in terms of the velocities at the six nodal points. The equilibrium conditions would be applied indirectly by means of a variational principle. This would involve minimising a potential-energy function with respect to the nodal point velocities, resulting in a set of linear equations for these unknowns.

Finite element methods can also be applied to problems of the biharmonic type described in Section 6.2.8. For example, two-dimensional slow viscous flow can be analysed using triangular elements of the type shown in Fig. 7.7. The velocity components in the x and y-directions can be expressed as linear functions of position within each element, and equilibrium conditions applied directly.

7.4.3 Three-dimensional elements   The finite element method can be applied to three-dimensional problems. Numerous element shapes are possible, the simplest being the tetrahedron with nodal points at the four corners. As with finite difference methods, the main limitation in the analysis of three-dimensional problems is one of cost. If sufficient numbers of elements and nodal points are used to give an acceptable degree of detail in the solution, the resulting set of linear equations may be prohibitively expensive to solve in terms of both computing time and storage requirements.

7.4.4 Comparison of finite element and finite difference methods   The main advantage of finite element methods over the finite difference methods presented in Chapter 6 is the much greater degree of geometric flexibility they offer. Using triangular two-dimensional elements, for example, arbitrary boundary shapes can be accommodated, provided they can be adequately approximated by a series of sufficiently short straight lines. The locations of the nodal points within the solution domain can also be chosen to suit the particular problem. Nodal points can be concentrated in regions where the values of the variables change rapidly, in order to maximise the accuracy of the solution obtained using a given number of points. Because of the geometric flexibility associated with finite elements, it is not generally necessary to employ a variety of co-ordinate systems. Most problems can be solved using cartesian co-ordinates. In contrast, finite difference methods usually employ co-ordinate systems whose axes are parallel to the boundaries of the solution domain.

    Whereas finite difference methods are based on the governing differential equations for the particular problem, finite element methods rely on either the direct application of equilibrium conditions, or the formation of an appropriate variational principle. The clear physical interpretation associated with the direct equilibrium analysis in finite element methods may be regarded as an asset from an engineering point of view. As shown in Section 7.4.1, however, the finite difference and finite element methods are fundamentally very similar, and under appropriate conditions result in identical sets of linear equations. Their orders of accuracy are the same for a given number of points in the solution domain.

7.5   Case Study: One-Dimensional Finite Element Analysis of Fluid Flow

This case study is concerned with the application of the analysis described in Section 7.4.1 to a particular problem. In addition to the velocity distribution, the volumetric flow rate is also required. As the lengths of the elements may vary, this is best obtained using the trapezoidal rule described in Section 4.1.1, which gives the flow rate per unit channel width as

$$Q = \int_0^H w \, dy = \sum_m \tfrac{1}{2} L_m (w_i + w_j) \tag{7.26}$$

7.5.1 Problem specification   A fluid having a viscosity of 10 kN s m^{-2} flows between plane parallel surfaces a distance 10 mm apart, due to a pressure gradient of 12 MN m^{-3}. The velocity distribution and volumetric flow rate are to be determined by the finite element method, using ten one-dimensional elements. The results are to be compared with the exact solution for this problem.

7.5.2 Solution   Figure 7.6c shows the arrangement of elements used. The nodal points are numbered in sequence to make the overall stiffness matrix tridiagonal. The numbering of the elements can be arbitrary, however, and the order selected is as shown.

Figure 7.8 shows a main program for solving this problem with the aid of the subprogram TRIDIA, shown in Fig. 6.5, for solving tridiagonal sets of linear equations. The main program is very similar to the one shown in Fig. 7.3 for the analysis of pin-jointed structures. Many of the variable names are the same, the main exception being that the array W is used to store the nodal point velocities (in place of D for the nodal point displacements). The arrays are dimensioned to allow for up to 50 elements and 51 nodal points. The variables PZ and VISCOS are used to store the pressure gradient and viscosity respectively.

As in Fig. 7.3, the coefficients of the stiffness matrix are computed for each element in turn, according to equation (7.18), and added to the corresponding coefficients of the overall stiffness matrix as indicated in equations (7.22). Note that the column number in the overall stiffness matrix is modified to conform to the notation for tridiagonal matrices shown in equations (6.26). The pressure forces at the nodal points are computed according to equations (7.20) and stored in EFORCE, before being added to the overall pressure force vector as indicated in equations (7.22).

The boundary conditions are applied by modifying the relevant equation when a nonzero value of the restraint type counter is detected. The equations are solved with the aid of TRIDIA, and the volumetric flow rate found using equation (7.26).

7.5.3 Results   Figure 7.9 shows the results obtained with the specified data, and the element arrangement shown in Fig. 7.6c. For example, element number 5 has as its nodes the points numbered 3 and 4, and has a length of 0·001 m. Of the eleven nodal points, only the first and last are restrained to give zero velocities on the boundaries. The velocity reaches a maximum of 0·015 m s^{-1} at the mid point of the channel, and the computed flow rate is 0·99 x 10^{-4} m^2 s^{-1}.

The exact solution for the velocity profile is

$$w = \tfrac{1}{2}P_z(y^2 - yH)/\mu \qquad (7.27)$$

```
C   PROGRAM TO FIND FLOW RATE IN ONE-DIMENSIONAL FLOW BY FINITE ELEMENTS.
C
        DIMENSION  F(51),STIFF(51,3),W(51),Y(51),NPI(50),NPJ(50),NREST(51)
        REAL  L(50),K(2,2)
   1    READ(5,51) NEL,PZ,VISCOS
  51    FORMAT(I5,2F10.0)
        IF(NEL.LE.0.OR.NEL.GT.50) STOP
        WRITE(6,61) PZ,VISCOS
  61    FORMAT(54H1ONE-DIMENSIONAL FINITE ELEMENT ANALYSIS OF FLUID FLOW /
       1   5H PZ =,E12.4,7H N/M**3,5X,11HVISCOSITY =,E12.4,9H N-S/M**2  )
C
C   INPUT THE NODAL POINT DATA.
        NNP=NEL+1
        READ(5,52) (I,NREST(I),Y(I),N=1,NNP)
  52    FORMAT(2I5,F10.0)
C
C   INPUT THE ELEMENT DATA.
        READ(5,53) (M,NPI(M),NPJ(M),N=1,NEL)
  53    FORMAT(3I5)
C
C   PREPARE TO SUM THE STIFFNESS COEFFICIENTS.
        DO 2 I=1,NNP
        F(I)=0.
        DO 2 N=1,3
   2    STIFF(I,N)=0.
        DO 4 M=1,NEL
C
C   FORM THE STIFFNESS MATRIX FOR EACH ELEMENT.
        I=NPI(M)
        J=NPJ(M)
        L(M)=ABS(Y(I)-Y(J))
        FACT=VISCOS/L(M)
        DO 3 IRE=1,2
        DO 3 ICE=1,2
        K(IRE,ICE)=FACT*(-1.)**(IRE+ICE)
C
C   ADD ELEMENT STIFFNESS TO OVERALL STIFFNESS.
        IF(IRE.EQ.1) IROW=I
        IF(IRE.EQ.2) IROW=J
        IF(ICE.EQ.1) N=I
        IF(ICE.EQ.2) N=J
        IF(N.LT.IROW) N=1
        IF(N.EQ.IROW) N=2
        IF(N.GT.IROW) N=3
   3    STIFF(IROW,N)=STIFF(IROW,N)+K(IRE,ICE)
C
C   SUM THE PRESSURE FORCES.
        EFORCE=-0.5*PZ*L(M)
        F(I)=F(I)+EFORCE
   4    F(J)=F(J)+EFORCE
C
C   APPLY THE BOUNDARY CONDITIONS AND SOLVE THE LINEAR EQUATIONS.
        DO 5 I=1,NNP
        IF(NREST(I).EQ.0) GO TO 5
        STIFF(I,1)=0.
        STIFF(I,2)=1.
        STIFF(I,3)=0.
        F(I)=0.
   5    CONTINUE
        CALL  TRIDIA(STIFF,F,W,51,NNP)
C
C   CALCULATE THE FLOW RATE.
        Q=0.
        DO 6 M=1,NEL
        I=NPI(M)
        J=NPJ(M)
   6    Q=Q+0.5*L(M)*(W(I)+W(J))
C
C   OUTPUT THE RESULTS.
        WRITE(6,62) (M,NPI(M),NPJ(M),L(M),M=1,NEL)
  62    FORMAT(28H0      M      I      J    LENGTH  / (1X,3I5,E12.4))
        WRITE(6,63) (I,NREST(I),Y(I),W(I),I=1,NNP)
  63    FORMAT(35H0      I    REST      Y            W        / (1X,2I5,2E12.4))
        WRITE(6,64) Q
  64    FORMAT(23H0VOLUMETRIC FLOW RATE =,E12.4,7H M**2/S)
        GO TO 1
        END
```

Fig. 7.8   Main program for one-dimensional finite element analysis of fluid flow

```
ONE-DIMENSIONAL FINITE ELEMENT ANALYSIS OF FLUID FLOW
PZ = -0.1200E+08 N/M**3        VISCOSITY =  0.1000E+05 N-S/M**2
        M     I     J     LENGTH
        1     1     2   0.1000E-02
        2    10    11   0.1000E-02
        3     2     3   0.1000E-02
        4     9    10   0.1000E-02
        5     3     4   0.1000E-02
        6     8     9   0.1000E-02
        7     4     5   0.1000E-02
        8     7     8   0.1000E-02
        9     5     6   0.1000E-02
       10     6     7   0.1000E-02

        I    REST      Y               W
        1     1     0.E+00          0.E+00
        2     0     0.1000E-02      0.5400E-02
        3     0     0.2000E-02      0.9600E-02
        4     0     0.3000E-02      0.1260E-01
        5     0     0.4000E-02      0.1440E-01
        6     0     0.5000E-02      0.1500E-01
        7     0     0.6000E-02      0.1440E-01
        8     0     0.7000E-02      0.1260E-01
        9     0     0.8000E-02      0.9600E-02
       10     0     0.9000E-02      0.5400E-02
       11     1     0.1000E-01      0.E+00

VOLUMETRIC FLOW RATE =   0.9900E-04 M**2/S
```

Fig. 7.9   Results from finite element program

and the maximum velocity (at $y = \frac{1}{2}H$) is

$$\hat{w} = -\frac{P_z H^2}{8\mu} = \frac{12 \times 10^6 \times 0.01^2}{8 \times 10^4} = 0.015 \text{ m s}^{-1}$$

The computed result agrees with this to at least four significant figures. The present finite element method is of the same order of accuracy as the finite difference method described in Section 6.3.1. The truncation error for the latter is given in equation (6.19), and is zero for a quadratic function of the form shown in equation (7.27). Consequently, for this particular problem there is no advantage to be gained by varying the lengths of the elements.

The exact solution for the volumetric flow rate is

$$Q = -\frac{P_z H^3}{12\mu} = \frac{12 \times 10^6 \times 0.01^3}{12 \times 10^4} = 10^{-4} \text{ m}^2\text{s}^{-1}$$

The error in the computed result is therefore 1 per cent, due to the method of integration rather than inaccuracies in the computed velocities.

While this case study provides only a very simple example of the application of finite element methods to continuum mechanics problems, it serves to emphasise the similarity with methods of structural analysis. It also provides an introduction to what are important methods of analysis for engineering problems.

8 Further Applications, and Classification of Problems

The main aims of this final chapter are to outline further applications for computers in engineering, and to consider the general classification of problems into mathematical and physical types. There are many applications which have not been considered here, notably those in the areas of management science and production engineering. Also, the emphasis in earlier chapters is on methods for analysing well-defined physical problems, rather than the application of such techniques to the design of engineering components and systems.

8.1 Applications to Design

As indicated in Section 2.1, problems of the design type are often solved by using a computer to explore a range of possible solutions, to which the design criteria can be applied. For example, in Section 7.2 the problem specification only called for displacements and forces in particular pin-jointed structures to be found. For design purposes it would be necessary to check that the displacements are acceptable, and that the forces in the elements are safe in terms of maximum permissible stress in the material and buckling of elements in compression. Changes in, say, the cross-sectional areas of the elements might have to be made to satisfy these criteria. An additional criterion might call for a structure of minimum weight or minimum cost. Optimisation problems of this type are discussed in Section 8.3.

For design problems involving changes in geometry of components or structures, the use of visual displays as described in Section 1.1 can be particularly helpful. For example, if a ship's hull is to be designed, the shape can be 'drawn' on the screen and then subjected to analyses to determine, say, the drag coefficient and the stiffness of the hull structure.

In addition to designing the components of an engineering system in terms of permissible stresses and acceptable stiffness it is often essential to consider the dynamic behaviour of the overall system. Many systems can be adequately represented by mass-and-spring models of the type analysed in Section 5.9, in order to estimate their natural frequencies. An energy method similar to that described in Section 4.3 can be applied to continuous structures. Natural frequencies should differ substantially from the frequencies of any loadings applied to the system, in order to avoid resonance which may result in gross displacements and failure. Section 5.7 provides an electrical example of the resonance phenomenon.

8.2   Management and Production Engineering

There are many applications for computers in management science and production engineering, although FORTRAN is often not the most appropriate language to use. For example, business accounting and the numerical control of machine tools each have their own specialised programming languages.

One of the most important areas of application is planning, which includes production scheduling, stock control, and the utilisation of labour. Typical planning problems might involve optimising the use of resources such as machine tools, or the minimisation of time and cost involved in plant maintenance. The scheduling of such complex activities is often analysed with the aid of the critical-path method. This serves to define which activities have the most direct effect on the overall time for the project, and how best to schedule all the activities in order to complete the project within a specified time and at minimum cost.

8.3   Optimisation

The ultimate objective of most types of engineering analysis is some form of optimisation. This applies to problems of both design and organisation. Although sophisticated mathematical techniques can be used in the optimisation process, the first priority is to decide on the optimisation criterion, the parameters that are to be varied, and the extent to which they can be varied.

One mathematical technique which is applicable to some optimisation problems is that of finding the maximum or minimum of a function by differentiating with respect to each of the variables in turn. This method was used in the curve-fitting process described in Section 3.3, and was mentioned again in Section 7.4.2 in connection with the variational formulation of finite element analyses. The required values of the variables are obtained by solving a set of simultaneous, though not necessarily linear, algebraic equations. Among other methods of optimisation are those which come under the headings of dynamic and linear programming.

8.4   Classification of Problems

The engineering problems examined in this book have been introduced under types of mathematical method of solution. In addition to summarising such methods, it is also useful to classify problems according to physical type.

8.4.1 Summary of mathematical methods   Most mathematical methods used in engineering analysis fall into one or more of the following three main categories.

(1) Processing of numerical data

Examples include the integration of experimental data (Section 4.2), interpolation of tables (Sections 3.5, 3.6 and 4.5), curve fitting (Sections 3.3, 3.4 and 4.5) and the application of other statistical techniques. Note that curve fitting by the method described in Section 3.3 involves solving a set of linear equations.

(2) Solution of differential equations

Differential equations are either ordinary (Chapter 4) or partial (Chapter 6), according to the number of independent variables. Methods of solution include the use of infinite series (Section 3.2), numerical integration (Sections 4.1, 4.2, 4.3, 6.6 and 7.5), step-by-step methods (Sections 4.4, 4.5, 4.6, 6.7 and 6.8), finite difference methods (Chapter 6) and finite element methods (Chapter 7). The choice of a particular method is influenced by the type of physical problem modelled by the differential equation.

(3) Solution of algebraic equations

Algebraic equations are either linear (Chapters 5, 6 and 7) or nonlinear (Sections 3.7, 3.8, 3.9 and 4.6). Nonlinear equations in more than one variable are often very difficult to solve, and have not been considered in this book. While the functional form of a nonlinear equation may be known explicitly (Sections 3.8 and 3.9), in some cases it can only be obtained numerically (Section 4.6). Sets of simultaneous linear equations may arise directly from the mathematical model (Sections 5.3, 5.4, 5.7, 5.9 and 7.2) or as a result of applying particular methods of numerical analysis, such as finite difference and finite element methods.

There is one type of computational technique that is often used in conjunction with various mathematical methods, namely the iterative technique. For example, nonlinear equations may be solved by the functional iteration (Sections 3.7.1 and 3.8), Newton–Raphson (Sections 3.7.2 and 3.9) or step-halving (Section 4.6) methods. Similarly, the Gauss–Seidel method for simultaneous linear equations (Sections 5.2.1, 5.4 and 6.6) is iterative, as is the correction process in the predictor-corrector method (Sections 4.4.2 and 4.5), and the method for finding the largest eigenvalue of a matrix (Sections 5.8.1 and 5.9). Although iterative techniques involve relatively large amounts of arithmetic, they offer two advantages for use in digital computations. As indicated in Sections 3.9.2 and 5.5.3, they cause minimal accumulations of roundoff errors, and because they involve the repetitive application of comparatively simple operations they are easy to program. The disadvantages of iterative techniques are that they may not converge, and that even if they do it is sometimes difficult to devise an adequate criterion to use in testing for convergence.

8.4.2 Types of physical problems   Engineering components and systems are usually represented for the purposes of analysis by models which are either discrete, or continuous in one or more space dimensions. The behaviour of either type may be time-dependent. For example, in Section 4.5 the motion of a railway train is represented by that of a discrete particle. Similarly, the models of oscillating electrical and mechanical systems considered in Sections 5.7 and 5.9 involve discrete inductances and capacitances, and masses and springs respectively. On the other hand, the model used for the buckling shaft problem considered in Sections 4.3 and 4.6 is continuous in one dimension, and in Chapter 6 most of the models are two-dimensional.

Most discrete models are of the 'lumped-parameter' type, where the behaviour of a continuous physical system is approximated by concentrating the variables into discrete lumps. Finite difference and finite element methods can also be regarded as methods of discretisation. The models for many systems should ideally be both three-dimensional and time-dependent. It is only with the aid of simplifying assumptions that the time dependence may be removed and the number of dimensions reduced. Although such simplifications are usually necessary to reduce the cost of computation to an acceptable level, they are often the cause of more serious errors than those introduced in the solution procedure. For example, by assuming a process to be steady, the possibility of unstable behaviour is excluded.

The categories discrete and continuous apply to the type of model and level of approximation used. The physical problems to which these models are applied generally fall into one of the following three classes.

(1) Equilibrium problems

These involve systems either at rest or in steady motion, including oscillations with constant amplitude. As shown in Section 6.1, equilibrium problems in continuous media are of the boundary-value type, and are associated with elliptic partial differential equations. Other examples include the structural analysis problem described in Section 7.2, and the electrical network problems described in sections 5.3 and 5.7.

(2) Eigenvalue problems

These may be regarded as types of equilibrium problems in which critical values of one or more parameters must be found, in addition to the corresponding configuration of the system. Examples include the determination of natural frequencies of oscillating systems (Section 5.9), and the loads required to buckle structures (Sections 4.3 and 4.6).

(3) Propagation problems

These are of the initial-value type in the sense that if a particular state of the system is known, its subsequent behaviour can be determined. Although this propagation is usually with respect to time (Sections 4.5 and 6.8), in some cases it may be with respect to a space co-ordinate (Section 6.2.6).

8.5 Problems Involving Nonlinear Materials

Virtually all of the problems considered in this book involve solid or fluid media which are linear. That is, their properties such as Young's modulus and viscosity are assumed to be constant. While this assumption is acceptable for many traditional engineering materials, it is not so for materials such as rubbers and plastics. Also, metals stressed beyond their elastic limits are nonlinear.

While methods of obtaining analytical solutions often cannot be applied to problems involving nonlinear materials, numerical methods can in principle be used with comparatively minor modifications. For example, suppose that the fluid involved in the channel-flow problem described in Section 6.6 is a molten polymeric material whose viscosity is a power-law function of the local shear rate, of the type discussed in Section 3.4. Although the type of problem is unchanged, the governing partial differential equation is nonlinear in the sense defined in connection with equation (6.1). Such a problem can be solved, however, by the finite difference method described in Sections 6.5 and 6.6. The viscosities at the grid points can be held constant during one or more cycles of the Gauss–Seidel process, before being updated according to the latest values of the local shear rates. A similar procedure can be used with an appropriate finite element method. It should be noted that such problems effectively involve sets of simultaneous nonlinear algebraic equations, for which direct methods of solution are inappropriate.

Another consequence of nonlinear material properties is that the equations governing a problem are not only nonlinear, but also 'coupled'. For example, in the fluid flow problem considered in Section 6.4, the velocity distribution was obtained independently of the temperature distribution. If the fluid viscosity is a function of temperature, however, the differential equations of flow and heat transfer must be solved simultaneously.

.6   The Influence of Computer Hardware

'he subject of computer hardware was introduced in Section 1.1, and its
ıfluence on engineering computations was briefly discussed in Section 2.2. This
ıfluence can now be reviewed in terms of the types of problems considered in
arlier chapters. The size of the fast core store is important in that it tends to
.mit the size or degree of complexity of the problems the computer can
onveniently be used to solve. Although backing stores are usually available, a
omputer with a larger fast core store is often preferred for solving larger
ıroblems. The speed of the arithmetic unit is also important in that it determines
he time and hence the cost of solving a problem. Since the costs for a given
mount of time on different machines reflect their relative speeds, a more
mportant consideration may be the maximum amount of time permitted for a
ingle job.

   Computer hardware limitations most commonly affect the solution of sets of
imultaneous linear equations. Computing times and storage requirements were
ompared for direct and iterative methods of solution in Sections 5.5.1 and
>.5.2. There it was shown that for large sparse sets of equations, the Gauss-Seidel
method requires fewer numbers to be stored than the gaussian elimination
method, and may be faster to compute (assuming that only the fast core store is
ısed). This superiority was confirmed in Section 6.6, for the finite difference
iolution of Poisson's equation. Taking this problem as an example,
ıpproximately n words of core store are required, n being the number of points
n the solution domain. Thus, a computer with a fast core store of 10 000 words
would only be convenient for a two-dimensional grid with less than 100 x 100
points, or a three-dimensional grid with less than about 21 x 21 x 21 points.
Similarly, if a grid of 300 x 300 points is required to obtain sufficient detail or
ıccuracy in the solution, a computer with a core store of at least 90 000 words
would be sought. If such a machine is not available, backing stores must be used.
The alternative direct elimination method of solution may then be more
efficient.

   With some programs, the amount of fast core store required can be reduced at
the expense of increased execution time. Instead of storing intermediate results
in a calculation, they can be recomputed when required. This and other aspects
of coding efficiency were considered in Section 2.6.1.

   A further hardware consideration is the time involved in input and output
operations, particularly the latter. As mentioned in Section 1.1, the speed of the
line printers may limit the capacity of the whole computer if the programs
involve excessive amounts of output. Ideally, the time required for printing the
results from a program should not exceed the total computing time involved in
compilation and execution.

8.7   Concluding Remarks

The advent of high speed electronic digital computers has broadened considerably the scope of analysis in all branches of engineering. Problems do not have to be simplified so far as to be amenable to the traditional methods of analytical solution. Nevertheless, the cost of using a computer to obtain numerical solutions is always an important consideration.

The correctness and accuracy of the results obtained with the aid of a computer are very dependent on the skill of the programmer, and particularly on his ability to test programs. To successfully program an engineering problem, he must have a sound knowledge of the problem, the mathematical model, the relevant numerical analysis, and the programming language.

The future of engineering analysis using computers lies in the direction of automating the design process, and optimising both engineering systems and methods of production.

Bibliography

Bickley, W. G. and Thompson, R. S. H. G. (1964). *Matrices: Their Meaning and Manipulation,* English Universities Press, London

British Transport Commission (1958)., *Bulletin No. 20, Performance and Efficiency Tests: Southern Region Modified Merchant Navy Class 3-Cyl. 4-6-2 Express Passenger Steam Locomotive No. 35020*

Conte, S. D. (1965). *Elementary Numerical Analysis*, McGraw-Hill, New York

Crandall, S. H. (1956). *Engineering Analysis*, McGraw-Hill, New York

Desai, C. S. and Abel, J. F. (1972). *Introduction to the Finite Element Method*, Van Nostrand-Reinhold, New York

Fenner, R. T. (1970). *Extruder Screw Design*, Iliffe, London

Ford, H. (1963). *Advanced Mechanics of Materials*, Longmans, London

Haywood, R. W. (1968). *Thermodynamic Tables in SI (Metric) Units*, University Press, Cambridge

Hoel, P. G. (1964). *Introduction to Mathematical Statistics*, 3rd edn, Wiley, New York

Isaacson, E. and Keller, H. B. (1966). *Analysis of Numerical Methods*, Wiley, New York

Kreitzberg, C. B. and Schneiderman, B. (1972). *The Elements of Fortran Style*, Harcourt Brace Jovanich, New York

McCracken, D. D. and Dorn, W. S. (1964). *Numerical Methods and Fortran Programming*, Wiley, New York

McCracken, D. D. (1972). *A Guide to Fortran IV Programming*, 2nd edn, Wiley, New York

Ralston, A. (1965). *A First Course in Numerical Analysis*, McGraw-Hill, New York

Schenck, H. (1960). *Heat Transfer Engineering*, Longmans, London

Timoshenko, S. P. and Gere, J. M. (1961). *Theory of Elastic Stability*, 2nd edn, McGraw-Hill, New York

Varga, R. S. (1962). *Matrix Iterative Analysis*, Prentice-Hall, London

Williams, P. W. (1972). *Numerical Computation*, Nelson, London

Zienkiewicz, O. C. (1971). *The Finite Element Method in Engineering Science*, McGraw-Hill, London

Index